W9-BDM-024

Down from the Mountain

Grand Prize winner of the Banff Mountain Book Competition
An Amazon Best Science Book of 2019
A *Literary Hub* Most Anticipated Title of 2019
An EcoWatch Best Book of April
A *Missoulian* Best Book of April
A *Chicago Review of Books* Best Book of April

"The two sides of Bryce Andrews — enlightened rancher and sensitive writer — appear to make a smooth fit . . . Precise and evocative prose."
— *Washington Post*

"Andrews's writing about wilderness is much like that of author Rick Bass, who displays both a healthy reverence for ecology and an easy way of talking about it. This story is not just about Andrews's shift from rancher to conservationist. It's an ode to wildness and wilderness in the form of grizzlies. It's about the tightrope bears walk between living in their mountainous territory, consuming pine nuts, army cutworm moths, and winterkill, [and] coming down the mountain to scavenge in human territory. It's about the resulting relationship between humans and grizzlies when they live in close proximity."
— *Outside*

"Andrews's wonderful *Down from the Mountain* is deeply informed by personal experience and made all the stronger by his compassion and measured thoughts. He outlines clearly the core of a major problem in the rural American West — the disagreement between large predatory animals and invasive modern settlers — without disrespect and without sentimentality. His book is welcome and impressive work."
— Barry Lopez

"In some of the clearest prose the state of Montana has produced, this high-octane story captures the marvel that is a grizzly giving birth in the high wild, follows her down into the human interface, and floods us with the heightened awareness and humbling unease we feel in the presence of *Ursus arctos*. When the hubris of man-unkind then threatens his protagonists, Andrews lays his life on the line in a sustained attempt to protect them, and the suspense of the telling comes to rival a great crime thriller. Rife with lyrical precision, firsthand know-how, ursine charisma, and a narrative jujitsu flip that places all empathy with his bears, *Down from the Mountain* is a one-of-a-kind triumph even here in the home of Doug Peacock and Douglas Chadwick."

> — David James Duncan, author of *The River Why* and
> *The Brothers K*

"Would that we had more nature writing like Andrews's fantastic second book, *Down from the Mountain*. Part biography of the Mission Valley in Montana, informed by the Blackfeet and Salish histories rooted there, it tells a moving modern tale of how ranchers and big predators overlap uneasily on that land today . . . *Down from the Mountain* eschews easy moral scrimmaging . . . A subtle and beautifully unexpected book . . . Readers hungry for yet another torch bearer to the ways of thinking of the wild that Barry Lopez and Leslie Marmon Silko made possible should look no further."

> — *Literary Hub*

"A cautionary tale of human-grizzly coexistence (or lack thereof). The book helps to illustrate the broader issues affecting grizzlies as their populations grow, pushing them closer and closer to humans."

> — *EcoWatch*

"The reader learns the history of the Salish land, and the varied new-comers to the valley and efforts to preserve and protect the grizzlies."
— *Missoulian,* "Montana Bookshelf"

"Beautifully written . . . Andrews conveys his passion for the West's landscape and inhabitants through his sensitive writing, which avoids either anthropomorphizing the wildlife or villainizing ordinary people . . . His book is a testament to his compassion."
— *BookPage*

"Andrews's writing reaches high peaks in *Down from the Mountain* . . . Andrews artfully describes the awe inspired by grizzlies . . . *Down from the Mountain*'s prose ranges in tone from graceful to elegiac to gripping."
— *Pacific Northwest Inlander*

"In stunning prose, as powerful as the grizzly itself, Andrews draws the reader into the mysterious lives of these bears. From deep in their pungent winter dens we emerge with them into the spring light, pad along forest trails, smell every molecule of wild and human. We are also the farmer, sweat-soaked, protecting the sweet corn. When these two worlds — bear and human — collide, all is unpredictable and precarious. *Down from the Mountain* will sear its beauty and sorrow into your soul. Required reading for all *Homo sapiens.*"
— Elisabeth Tova Bailey, author of
The Sound of a Wild Snail Eating

"The fluidity of Andrews's imagination and the reality of his time on the ground with people and bears make this book a piece of true history. He is not a journalist, but a participant with skin in the game who happens to be an excellent writer. Putting up fence along the porous line between humans and big, foraging bears, Andrews is the one you'd want telling this story."

— Craig Childs, author of *The Animal Dialogues* and *Atlas of a Lost World*

"In *Down from the Mountain*, Andrews walks the harrowing line between wilderness and civilization — as in literally walks it, recounting his own efforts to keep a space in the world for the untamed creatures that remind us who we are in the first place. Writing with a keen empathy for both the great grizzlies of Montana's Mission Mountains and the farmers and wildlife officers coping in the valley below, this book is by turns heartbreaking and hopeful, even while it zings along with the high-stakes pace of a thriller. It's as true as it gets."

— Malcolm Brooks, author of *Painted Horses*

"Returning home from ten days in the backcountry, I devoured this fabulous and feral book in a single sitting and found myself utterly immersed in the 'unforgiving arcadia' that is our vanishing West. 'Bears are made of the same dust as we,' John Muir reminds us, and this marvelous narrative, even in passages devoid of humans — perhaps especially in those passages — draws us into communion with these uncompromisingly powerful wild creatures, to the heartbreaking consequences of our inevitable encounters with them, and to one man's profound compassion for them. For two decades as a hunter, angler, and hiker, I've traversed the very country described herein. I have never seen it with such sustained clarity as through the vital lens of Bryce Andrews's luminous prose."

— Chris Dombrowski, author of *Body of Water*

"A thoughtful story of bears, humans, and their tragic interactions . . . A gem of environmental writing fitting alongside the work of Doug Peacock, Roger Caras, and other champions of wildlife and wild land."   — *Kirkus Reviews*, starred review

"Andrews, a conservationist and rancher in Montana's Mission Valley, examines dramatic changes in the local bear population, which once 'lived a grizzly's solitary life' but now show up regularly near human dwellings searching for food, in his compassionate study . . . Andrews's well-written cautionary tale leaves readers with the sobering message that humans must [modify their behavior] if they are to be responsible stewards of nature."

— *Publishers Weekly*

"This fascinating, well-researched, and lyrical memoir will appeal to conservationists, those curious about large predators, and readers who relish stories of the West."   — *Library Journal*

"[A] lyrical exploration of an attempt to accommodate two disparate goals — the dairy farmer's need for the corn to feed his cattle and the grizzly's need to eat and fatten up during the short Montana summer . . . Andrews's empathic writing turns Millie's story into the embodiment of modern compromise with apex predators."

— *Booklist*

Down
from the
Mountain

Down
from the
Mountain

*The Life and Death of
a Grizzly Bear*

Bryce
Andrews

MARINER BOOKS
HOUGHTON MIFFLIN HARCOURT
Boston ▲ New York

First Mariner Books edition 2020
Copyright © 2019 by Bryce Andrews
Reading Group Guide copyright © 2019 by Houghton Mifflin
Harcourt

All photographs are credited to the author with the exception of
those on pages 148 and 214–15 courtesy of Colleen Chartier and on
page 186 used by permission of Shannon Clairmont, CSKT Wildlife
Biologist.

For information about permission to reproduce selections
from this book, write to trade.permissions@hmhco.com or to
Permissions, Houghton Mifflin Harcourt Publishing Company,
3 Park Avenue, 19th Floor, New York, New York 10016.

hmhbooks.com

Library of Congress Cataloging-in-Publication Data
Names: Andrews, Bryce, author.
Title: Down from the mountain : the life and death of a grizzly bear
/ Bryce Andrews.
Description: Boston : Houghton Mifflin Harcourt, 2019.
Identifiers: LCCN 2018033157 (print) | LCCN 2018048021 (ebook) |
ISBN 9781328972477 (ebook) | ISBN 9781328972453 (hardcover) |
ISBN 9780358299271 (pbk.)
Subjects: LCSH: Grizzly bear—Montana. | Human-animal
relationships—Montana.
Classification: LCC QL737.C27 (ebook) | LCC QL737.C27 A536 2019
(print) | DDC
599.78409786—dc23
LC record available at https://lccn.loc.gov/2018033157

Book design by Greta D. Sibley

Printed in the United States of America
DOC 10 9 8 7 6 5 4 3 2 1

For two exceptional animals

CONTENTS

Down
from the
Mountain

1

The
Valley

SHEER PEAKS mark the eastern edge of Montana's Mission Valley. Gray Wolf, the southernmost, is a fist. East and West Saint Mary's are shoulders without a head between them. Kakashe is a rampart, with higher pinnacles overtopping it from behind like breakers cresting a seawall. All are made of gray stone, with fissures, walls, and drops that give pause to the most intrepid climbers. They are quick to gather snow, slow in losing it through summer.

The mountains, which form the eastern margin of the Flathead Indian Reservation, shelter a healthy population of grizzly bears. As fall gives way to winter, these bears climb. Disappearing into well-hidden dens, they wait for spring.

Up there in February, it looks like nothing is passing but time, wind, and occasional ravens, but grizzlies are busy underground. In secret, deep-drifted nooks, they breathe and stir. Sows give birth in darkness.

Sometime around the year 2000 — it might have been a winter or two before — in a den smelling of earth and ursine rancidity, a cub was born. She probably had a sibling or two, as grizzlies are mostly born in pairs or trios. Until spring came, the writhing of her littermates and her mother's warmth were her whole world.

Though much is now known about that cub—her life, movements, and the circumstances of her end—no one can say precisely where or when she was born. It is enough to say that she was born in the high country, into a den that no human found or entered. In this, she was like most bears that have been born in the Missions since the recession of Pleistocene ice revealed mountains to the sun.

Emerging into spring, she was shown by a careful mother how to get about in the mountains. She learned which things were to be eaten and which were dangerous. In time, she descended with her mother across ridges and avalanche chutes toward the valley floor. Down in that settled, domesticated landscape, she smelled and saw human beings for the first time. She learned to be wary of gravel roads and highways and to move discreetly among farms and the scattered houses of rural subdivisions.

She slept one more winter in her mother's den, woke, and made the seasonal round as a yearling. Then, breaking with her family, she went out alone. She grew into adulthood, weighing nearly five hundred pounds and measuring three feet tall at the shoulder. Rising on her hind legs to pluck apples from a tree, she could reach higher than most people. Her forepaws were wide and black padded, and they hardened as she went about mapping the smells, contours, and hazards of her home range. She lived a grizzly's solitary life, and if she was seen at all by the men and women who lived in the valley, it was as a disappearing flash or a shadow against the night.

IN 2002, in the blue light that follows dusk in late summer, she left off foraging and walked down from the foothills into a tangled aspen grove. Crossing Millie's Woods—the copse for which she'd soon be named—the bear came to a place where she could see farmstead lights spread across the floor of the Mission Valley, glittering like shards of a bottle dropped from a great height.

While darkness thickened, she shuffled along a well-worn game trail, giving a generous berth to barnyards and houses. Hearing the faraway barking of dogs, she kept to the low ground of potholes and sloughs.

She came in search of ripening apples and the chokecherries weighing down the branches along the banks of irrigation canals. Descending step by cautious step, keeping pace with the night, she left the Mission Range behind. In doing so, she walked out of a wilderness that has remained essentially unchanged since the end of the last ice age, and into an unforgiving arcadia. The primeval valley — the fertile, deep soil that had succored native people and grizzly bears for thousands of years — was hidden by roads, power lines, prefabricated ranch-style houses, tilled fields, and uncountable miles of barbwire fence.

Cars and trucks hurtled day and night along Highway 93. Except for the timbered corridors along streams, the land was settled, cleared, and cropped. There were pastures and hayfields, gardens and chicken coops, pigsties, grain bins, and trash piles — all manner of things that could lead a bear into conflict with people or livestock, and therefore to ruin. The farms were small by Montana standards, with most holdings measuring between twenty and eighty acres. The landscape was not paved over or entirely ruined for a grizzly bear's purposes, as parts of the Missoula and Bitterroot valleys are, but it was a difficult, dangerous place to survive.

She passed near Schock's Mission View Dairy, a family operation with a ramshackle milking parlor and a wide yellow Quonset barn. Crossing behind the buildings, she heard the thick-voiced lowing of cows. She kept clear of it all, padding northward toward the shelter of a brushy stream. Nearing East Post Creek Road, she came to where the Schock family had planted their first field of silage corn — a crop that would be harvested to feed dairy cows through winter. Crossing gravel, leaving long-clawed tracks in the borrow ditches, she slipped in among the stalks.

Corn closed around on all sides, stillness descended, and broad leaves hissed across her shoulders. Ripe ears tapped her muzzle. With the fantastic acuity of a bear's nose, she smelled a faint sweetness all around. Instead of shambling ahead toward the half-wild, overgrown orchards along Post Creek, she stopped. Stretching out a forepaw, the sow snapped off a cornstalk at the base. Sitting back on her haunches, laboring delicately with paws strong enough to break the spine of an elk, she brought an ear to her mouth and chewed.

She tasted pulped kernels, husk, and silk. White-gold liquid dripped from her lips. With three-inch-long claws working as precisely as forceps, she husked another ear. The corn was as endless as her hunger, and under the pale moon, she gorged to the cusp of bursting. When she could hold no more, she rested. The field seemed a good, safe place at night. Though she retreated to the mountains before daylight, she returned again at dusk.

Every year, a bear must solve a simple but unforgiving problem: Coming out of hibernation as thin as a rail, each grizzly must gain enough weight in the months between April and November to last through winter. From a bear's perspective, the cornfield offered a perfect solution: an infinite, carbohydrate-rich source of food that ripened just as summer became fall. Having discovered the crop, the sow could no more desert it than stop breathing. There were other sources of food — apples, plums, and accumulations of insects high in the peaks — but she always returned to the corn. Nothing else was as sure or as plentiful. She frequented the field, and other grizzlies joined her. Yearlings, adrift from their mothers, blundered across the crop. Ponderous full-grown boars lumbered into the field and began to eat. Sows arrived with their new cubs, sought secluded corners, and fed.

Night after night, their numbers grew. The grizzlies toppled stalks, plucked ears, and bolted down kernels in an ecstasy of consumption. They could not have done otherwise. At the end of each summer, a bear's appetite turns desperate and insatiable. Hyperphagia is the mania's proper name, and it grips the animals tightly. The unique hunger that grows in a human stomach with the onset of cold — that ravenous demand for rich stew and potatoes — can only be a shade of what a bear feels. Theirs is a mortal compulsion to feed, exacerbated by shortening days.

The sow stayed until the crop was harvested. She left the field swathed in subcutaneous fat, which kept her warm through winter. It was no wonder she returned the following year and that other bears did, too.

In Schock's cornfield, grizzlies learned to gather in concentrations that are rare in the Northern Rockies. They massed as Alaskan bears do around salmon streams when the fish are running. Hunger conquered their solitary natures, and the bears ate together, widening the circles of flattened stalks.

Cubs learned the behavior from their mothers, and adults found their way to the crop with astonishing rapidity. According to hair samples from rub trees and genetic tests performed on dead and anesthetized bears, there are at least seventy-five grizzlies living in the Mission Range. After the corn had been growing for a few years, up to a quarter of those bears were feeding in Schock's cornfield between August and mid-September.

Two things followed from this: First, Greg Schock lost a lot of corn — thousands of dollars' worth, enough to threaten the solvency of his farm. Second, the grizzlies began eschewing their traditional alpine sources of late-summer food. They no longer climbed to the stone fields on McDonald Peak to eat colonies of army cutworm moths or searched the high forests for whitebark-pine-nut caches laid in by red squirrels. Instead, enraptured by the corn, they stayed in the low country. This fattened the bears up marvelously, but it also kept them close to humans during their hungriest, most aggressive season, putting them at odds with Schock and other farmers and increasing the likelihood of confrontations.

STACY COURVILLE, a wildlife biologist for the Confederated Salish and Kootenai Tribes, is a sturdy man. He moves powerfully and somewhat heavily, like a bear. He was new to the task of trapping grizzlies in 2006 when he started working on a region-wide study of bear genetics, migration patterns, and habitat use; but I suspect that even then he was steady. Then, as now, he was good in a tight spot.

After an education at the University of Montana and a stint with the United States Forest Service, Stacy had taken a job on the Flathead Reservation in the Mission Valley. The work meant coming home; he is an enrolled member of the tribes and was raised in the small town of St. Ignatius, just five miles south of the Schock dairy and its corn. Having grown up fishing, hunting, and walking in the Missions, he knew how many grizzlies bedded in Millie's Woods and had come to appreciate the power and swiftness of bears. He did not enter that grove without good reason, caution, and a gun. Still, as he drove into Millie's Woods in May 2006, accompanied by an older biologist, Stacy must have been nervous about the task ahead.

Among the aspens, the light turned false. Sieved through leaves, the sun's glow shifted and flashed on white bark, dying on the mulch of the forest floor. The truck passed through contrasting pools of shadow.

Coming to the sow, they found her anchored to the trunk of a spruce tree by a cable snare wrapped around her forepaw. She waited, smacking her jaws and huffing while Stacy estimated her weight, drew a measure of Telazol, and took up the tranquilizer gun. The sow flinched when the dart hit her rump.

They waited. She swayed but did not fall. In those days, the tribal biologists used syringes with adhesive collars. Those darts often shook loose when an animal struggled, and it was hard to know how much tranquilizer had been delivered. The bear stood her ground, snarling with fear and rage. The men wondered if something had gone wrong with the shot, or if the drug was bad, or if the bear had a supernatural hold on consciousness.

"Again?" they asked each other.

Another dart struck home, and she dropped in a heap.

The sow's breathing was shallow and quick. Her heart raced, but though she had been darted twice, she never descended to the limp oblivion of the surgical plane. As they took samples of hair and blood, estimated her age at around seven years based on the condition of her teeth, weighed her in a sling, measured her paws, and fitted the leather strap of a GPS collar around her neck, the sow's body went on working as hard as if she were running a marathon. Watching the frantic panting of her rib cage, they worried that her overstrained heart might fail.

STACY MADE a phone call, and a second vehicle crept into the woods. Climbing down from their ranch rig, Greg Schock and his daughter, Emily, admired the bear's bulk, the luster of her fur, and the long wheaten knives of her claws. Greg liked the grizzlies, and though he wished that they had never found his corn, he bore the animals no ill will. Emily had been doing a project on them, using motion-sensitive trail cameras to capture images of the animals that crossed her family's pastures at night.

As a rancher and a dairyman, Greg knew animals well. Observing the sow's condition, he must have thought, *Good, healthy bear*. His practiced eye would have seen that she was lactating, and he would have known that cubs, though they remained unseen, were nearby. Standing beside his daughter, he watched the animal's labored panting.

Stacy called the bear Millie, after the woods. The woods, in turn, were named for Millie Morin, a pint-size indomitable farmer who had made her home on the edge of the timber for seventy years.

The biologists made final adjustments to the sow's GPS collar. By 2006, that technology had given researchers a new perspective on the lives of grizzlies. A GPS unit, unlike the radio transmitters that had been used in wildlife studies for decades, recorded an animal's position at set intervals until its batteries ran low. If all went well, Millie's collar would drop from her neck on a predetermined day and broadcast a signal to Stacy, who would track the unit down and learn precisely where the bear had been.

As Stacy finished his work, Millie paused in her breathing. The biologists, the dairy farmer, and his daughter stood motionless under the spreading aspens for a handful of tense moments until the sow's chest began heaving again.

They lifted the bear into a culvert trap—a long steel cylinder with sliding gates at either end—administered agents of revival, retreated to the safety of their vehicle, and waited. When the sow rose swaying, they let her go. Millie vanished into the thick undergrowth of the woods, and to Stacy, it seemed that they'd come close to killing her.

2

Newcomers

I N 2002, when corn was first planted on the Schock dairy and bears began finding it, I was a college sophomore in Walla Walla, Washington. By 2006, when Stacy Courville trapped and named Millie, I had graduated and was headed for a ranch job in Montana's Madison Valley. On the Sun Ranch, in a valley near Yellowstone National Park, I had my first taste of what it was to make my living as a rancher and live among grizzly bears.

That work, begun humbly as a summer hand, became my livelihood for a decade. I devoted myself completely, wanting very much to be transformed by agriculture. I meant to lose the softness of an urban Seattle boyhood and be beaten by labor into the shape of a man.

I worked on large ranches in the mountains adjacent to Yellowstone Park. In autumn, it was often my lot to ride out before first light in search of stray cattle and trespassing hunters. The images, sounds, and feelings of that task have not left me. I remember what can be heard once the eyes give up on seeing, the cold that prevails before the slow brightening, how much a horse knows about the shrouded world. But describing the brittleness of frosted grass, the way my lungs ached and vibrated in tune with an empty sky, or how easy it was to lose faith in the sun's coming falls short of describing that hour and the work.

To my way of thinking, there are two kinds of mountains in Montana: those that still contain grizzlies and those that have lost them. There are large carnivores all over the Rocky Mountain West, and some of them are fearsome enough. Anyone who has seen a pack of wolves run an elk to exhaustion or witnessed the smooth, dangerous approach of a mountain lion will agree. A black bear, too, is a creature worth respecting. But nothing else is like a grizzly.

This seemed particularly true when I was on horseback in the hours before morning. The threat of bears loomed large on my predawn rides because predators love and own the crepuscular hours. As I went out from the barn, with the horse jigging and only halfway sold on the idea of walking alone, darkness would press in from all sides.

On one ranch where I worked and grizzlies were particularly numerous, the land was shaped like an upturned hand, with timbered fingers running down from the peaks to the valley floor. Making my rounds on that place meant crossing many creeks, and there were always noises in the brush.

Stirrings pricked the horse's ears and set his muscles tightening. That tautness passed up through my legs and gripped my heart. As if the stream of time had clotted, everything slowed while we strained to discover what the hidden creature was. Those moments were terrible and united me in purpose with my mount. We bent our attention to sounds, dim shapes, and hints on the wind. The horse knew more than I did, and when relief entered his body, it washed like a tide over me.

On such mornings, I came to understand how the proximity of

grizzlies changes a person. I tasted fear, which burns the tongue's tip like copper, and felt my body knot.

WHEN I worked as a rancher, such encounters were commonplace. Bears left tracks in mud and snow, and I learned to be cautious. Wolves howled at dusk, and I headed out to check on my herds. Whitetail deer haunted the edges of every hayfield, and elk came down from the heights ahead of the first serious snowstorms, whistling and chirping to one another like birds.

Wild animals were everywhere around me, but they were seldom at the heart of my work, which was the care and feeding of livestock. I was good at ranching and gave myself wholly to it, rearranging my life and morality in accordance with an ancient bargain.

The proper word for it, "husbandry," has often tempted me to imagine a marriage ceremony with the grazier opposite his cow. The vows, as I understand them, could be articulated as follows:

Do you, herdsman, accept responsibility for these creatures? Will you wake in the night when they bawl? Protect them from harm? Clear their path and mend the destruction following in their wake?

And will you, cow, stay gentle? Will you suffer an occasional touch, remaining infantile your whole life through? Will you grow fat on simple rations and surrender your progeny? When the time comes, will you stand easy in the kill chute? Without complaint or much comprehension, will you die?

RISING FROM the ranks of hired help, I managed ranches. By 2013, I had a small spread of my own, which I shared with a friend named Bart. It feels strange to call the Oxbow Cattle Company a ranch because we owned almost nothing, and I think of ranchers as people who own a lot. We grazed cattle on the edge of Missoula, on three thousand leased acres of weedy, unfenced, disused land. The whole business was cobbled together from borrowed pieces and had an improvisational beauty that I loved.

The heart of the operation was this: We sought out ranches farther up the Clark Fork River and bought from them heifers that had failed to deliver calves. We fattened these animals — a class of cattle called "heiferettes" by stockmen — on the lush grass that grows beside the lower Bitterroot River, selling their hormone-and-antibiotic-free meat locally.

Oxbow Cattle Company LLC came into being in January. The business cracked my heart open eight months later on a Wednesday in late August. I know it was a Wednesday because we had to keep a steady stream of meat coming to the supermarkets and restaurants in town, and Wednesday was hauling day. In the morning, Bart and I gathered the herd and sorted off four of the fattest animals. Because we handled our cattle calmly, the cows stepped easily into our trailer. Latching the door, I headed for the butcher's, leaving Bart to attend to some business in town.

I slid down the highway in perfect comfort — Bart's truck was the only newish piece of equipment that Oxbow owned — listening to music and feeling the rig shimmy with the movement of the livestock. The day was brilliant and warming. The Clark Fork River ran through blue shadows.

The butcher's place was small, and it sat within a stone's throw of the highway. Up front was a counter, in the middle was the cutting floor, and at the back were rails where carcasses hung in fat-sheathed rows. Out behind the building was a series of holding pens that ranked among my least favorite places in the world.

The corrals were small and high sided, with a sliding guillotine gate and a loading chute welded from heavy-gauge pipe. Built from arena panels and horizontal bands of freeway guardrail, the walls were constructed with a single goal in mind: that an animal, no matter what species and no matter how desperate, would never escape. The butcher once told me that before they added the up-permost rail — it hung at least ten feet off the ground — a bison had somehow clambered out.

"Had to shoot it on the road," he said. "Pain in the ass."

I backed the trailer up until its bumper kissed the load-out. Set-ting the gates in the pens, I could not help but notice the reek of old blood in the air. The stink was always present, but that day it was stronger. My cattle seemed to agree. When I slid back the door, they balked and stared wide-eyed.

Walking to the front end of the trailer, I climbed onto the run-ning board and peered through slatted metal. The heiferettes bunched together, their hindquarters tight against the gooseneck's steel.

"Come on," I said. "Out of there."

Nothing moved. They watched the open door's bright rectangle.

"Sk-sk-sk!" I chided. "Come on now. Out!"

Not one of them stirred.

"Hey!" I yelled, walking back, sticking my hand into the trailer and slapping the nearest rump. *"Hey-hey-hey-hey-hey!"*

Nothing. The cattle held as still as statues. There was no noise but their huffing breath.

In the end, I had to force them out, climbing through the man door in the side of the trailer, whooping and swatting. The little herd moved haltingly toward the opening at the back end of the trailer, regarding its threshold as if it were a cliff. I pressed my cause, yelling and slapping, full of unexpected frustration. When one of them plunged into the light, the rest followed on her heels.

As quickly as I could, I lowered the gate into place. Within the bars, my cattle, which were so calm in the pasture that I could walk up and touch their noses, were destroyed by fear. Two huddled in the pen's far corner, pressed so tightly together that they seemed to be a single beast. Another stood centered in her grim world. She dropped her head, blew through her nostrils, and pawed the cement floor. But it was the last cow that did me in. She went raving mad.

She made half a dozen circuits of the pen, spurting ahead like a spooked trout circling a pool. Each lap was faster than the last, with her hooves skittering over the grooved concrete.

"Easy, now," I said to her. "Easy."

Cutting a scrabbling turn, she ran headfirst at the fence, hitting it in full stride three paces from me. The guardrail flexed and rang, and I could see the impact rippling through her neck and across her well-fed belly. When her legs began to buckle, I thought that she had broken her spine.

She stayed up, unsteadily finding her balance and facing the bars. Charging again, she leaped with front legs outspread. One front hoof lodged tightly in a higher rail and she hung for a moment, moaning. In the end, she shook free, fell back heavily, and stood still. Spilling bright blood from one nostril, she offered the world an abject bawl and fixed me with a constant gaze.

An unhinged charge, a cry, and a stare. Those things knifed to my center and I thought: *She knows. Even if those others don't, she sees what's coming.*

I remember this very well: While I sat thinking about the cattle and their predicament, and how many times I had brought an animal to this same end — the heifer didn't blink. She watched me, and her steadiness was a kind of accusation.

What could I have done? The thing seemed sealed and finished. I knew that there was nowhere else to take the heifers if I wanted to make my living as a rancher. I told them that they had been good cows and that I was sorry. One of the butcher's hired men, coming out through the back door, said: "What's that, bud? You say something?"

"No," I answered. "Just dropping off a load."

Afterward, I never hauled cattle in peace. Each time I took to the highway, particularly when I did it without Bart's unflappable company, I harbored visions of pulling over in a lonely, unfenced field; swinging the door open; and freeing the cows. I daydreamed that prison break half a dozen times, but even as a dream, it wouldn't work. The cows stood dumbly in the field, waiting to be herded toward the next good place. I knew that if I left them there somebody would come along, call them strays, and cart them off for his own use. I understood that there was no other course for my heiferettes or any other cattle. They were born to die.

Sometimes I would take my cows to the sale yard, drop them off, and feel hollow afterward. Sometimes I felt nothing at all, which led me to believe that the essential gearing of my soul had been worn out by the task of turning animals into meat.

I could say a lot about the harder parts of ranching: that each person should kill no more than his fair share and leading cattle to die in great numbers without manifest remorse will leave him stunted and incapable in certain ways. That might be true, or it might just be true for me. Of this I can be certain: I recognized a void and a canker, and I was certain that it had something to do with killing. I suspected, too, that making amends would require saving the lives of animals.

I let Oxbow go, selling my half interest in the company to Bart. For the first time in years, I had no herd to care for or plans to find one. I was briefly adrift, then profoundly relieved as I came to understand that managing a ranch was not a prerequisite for happiness.

I worked and lived in town and fell in love with a dark-haired, joyful woman named Gillian. We walked through the city's greener neighborhoods at dusk, talking and glancing in at families through lit windows. Through the latter part of 2014, we began to build a life together in Missoula.

Though I was comfortable and pleased, something rankled. It first manifested as a subtle unease and a difficulty sleeping when streetlights split our curtains. Then it was a simmering discontent and a lack of patience with strangers in the street. I found myself jogging farther and farther on the trails and logging roads outside of town, with my eyes straying to the dark-thicketed mountains. Wild creatures were moving up there, I knew, and I missed being near them.

By the end of summer 2015, I understood that I could not sustain an urban life. As Gillian's efforts to earn a master's degree from the University of Montana's environmental studies program promised to keep us in Missoula for at least two years, I resolved to find a job that would send me into the hills.

That desire simmered through fall. It had reached a rolling boil by the time an old friend of mine, Steve Primm, approached me about working for a nonprofit conservation group he had started some years earlier.

People and Carnivores is an aptly named small organization based in Bozeman, Montana. When Steve and I first talked, the group had a staff of four working to mitigate the conflicts that arise when ranchers, farmers, hunters, and recreationists share landscapes with large predators. As the organization's conservation director, Steve was responsible for all that went on in the field. Listening to him talk about the group's projects — fencing efforts to keep bears away from barnyards, herding programs to minimize wolf conflicts in southwest Montana, food storage poles to keep hunters and hikers safe in the backcountry—I liked the sound of the work and the fact that it would carry me far from town. Steve, in turn, thought years of raising cattle would help me relate to and work with ranchers who were losing their livestock to wild predators.

I spoke with Lisa Upson, the group's executive director, and in short order had agreed to expand their efforts into the valleys around Missoula. We talked at length about how and where to begin, and I think that it was Steve who finally said: "What about the Flathead Reservation? They have a lot of grizzlies up in the Mission Valley."

I signed on with People and Carnivores in the first days of 2016, entering the field of large-carnivore conservation at nearly the same time that a pair of cubs began growing inside the sow that Stacy Courville had trapped and named Millie.

Though Mission Valley grizzlies breed in May or June, the development of fertilized eggs stalls after the creation of a small bundle of cells. Life slows down, and the embryo enters something akin to hibernation, remaining viable and unchanged through summer and fall. Pregnancy proceeds in midwinter, and only if a sow enters her den well-fed and healthy.

Millie was both, and as she slept through the first days of 2016, her cubs grew. They were born in February into the year's dark basement. Blind and sparsely haired, weighing twelve ounces each, they slipped out of her into a second womb of musty earth and crooked roots. They made plaintive noises, scrabbling at hard-packed soil until they found thick fur, digging for nipples as big as their heads. Millie did not fully rouse, though she shifted to aid their efforts. She unfurled herself and gave suck, presenting her skin to the cold so the cubs might live.

In the beginning, her newborns would have fit in the palm of a hand. A person, if he was lunatic enough to crawl into the den, could have strangled the cubs with a noose made of his thumb and forefinger. Such is the vulnerability of week-old grizzlies. Like us, they come into the world unfinished.

They grew in the dark. Mouse-brown fur covered their strangely human bodies. Their eyes opened, seeing nothing for a time, then spring's white light.

Leaving the den in early April, the cubs weighed around fifteen pounds. They walked through the timber of a north-facing slope. Already they had developed the shape of grizzlies: humps, long claws, and their mother's compact round ears. Coming to a ridgeline, they saw clear sky above jagged stone and felt the sun's warmth pour over them. Millie pulled at the wind with her wide black nose. She set a course for new growth and carrion, and the cubs followed.

CERTAIN BLESSED spots on Montana's western rampart, the Bitterroot and the Flathead valleys among them, remind me of coastal Washington. This lush, plentiful inflection is nowhere stronger than in the Mission Valley, where cattails choke the sloughs and fruit trees sprout unbidden. In the mountains grow healthy, if isolated, groves of red cedar. Bracken ferns sprout in the shelter of their boughs, rising plumelike from the duff.

A testament to the place's fertility: Years ago, I traded for a horse from the Mission. That gelding was a pretty, pedigreed descendant of a renowned paint stallion. When I picked him up, he stood belly deep in feed with his coat shining like brushed copper. I brought him home to the dry, thin-soiled valley where I worked, and over the subsequent months, I watched him waste away. I did everything that I could: bought supplements and hired specialist veterinarians. I heaped flakes of the best alfalfa hay into his manger, but he still diminished. In the end, he fell to skin and bones while the ranch's other horses flourished. The process moved quickly at its end, and by the time I had exhausted all my options, it was too late to take him home. When the vet put him down, it was an unmistakable mercy.

The tests said that his kidneys had failed and that his guts were full of swallowed sand. Privately, I supposed that the horse could not stand his new surroundings. Having lived with him for nearly a year, I knew what the stones did to his hooves. Watching him in the pasture, I saw that he was loath to chew on wire-hard, sun-dried grasses.

Of course, the Mission Valley's climate may not truly be called

mild. Cedars grow only in rare damp nooks, with expanses of ponderosa pine and Douglas fir covering the ridges between them.
Winters drop into the killing temperatures below zero, and the
wind, screaming from the north, will not let a person forget that
he stands high on the continent's spine. Therein lies the lure of the
Mission Valley and a large part of why I love the place so well: It
is a transitional country—fertile as often as it is unyielding, warm
enough to grow late-summer peaches and cold enough for weight-
less snow—bridging the Pacific Northwest of my youth with the
Rocky Mountains where I make my living.

THE MISSION'S bitter wind recalls the deep past. Twelve thousand years ago, at the end of the last glaciation, the valley was a
blue-white river of ice stretching northward to Canada. The rock
beneath that ice was ancient and mud cracked, exhumed from the
basement of the world. Gray-green in places, ruddy elsewhere, it
was deposited in still water some 1.4 billion years ago, compressed
for eons, and coughed up by a continental collision. The stone
shatters easily, breaking away in two-inch-thick slabs with sinuous
ripples and the pinprick marks of droplets. There are few fossils.
Beyond the occasional swirl of algae, life was too new to leave a
mark.

Glaciers rasped the valley into shape, scouring out Flathead
Lake, the largest naturally occurring body of fresh water west of
the Mississippi, and sharpening peaks into tricornered spearheads.
Abandoning its southward march as the climate warmed, the ice
left a moonscape of eskers and moraines on the valley floor. Li-
chens, mosses, and smaller plants spread onto that barren zone,
sucking meltwater and dying in their time, building humus.

The grizzly was there, but she was unremarkable among the
Pleistocene menagerie. She was a diminutive sister to *Arctodus*,

a two-thousand-pound short-faced bear. Walking belly-high to a ground sloth and knee-high to a mammoth, the grizzly was only a medium-size predator — a welterweight in a land of heavies. Like today's coyote, she knew what it was to hunt and be hunted. She went about in the ancient world as we now pass through the wildest corners of the modern one: with eyes peeled and her heart often in her throat.

HUMAN HUNTERS moved north on the heels of the receding ice, coming into the valley when the land was yet raw and studded with erratics. Only the first scrim of vegetation covered bedrock, and generations of men and women watched the river sort cobble from sand. They walked among new willow shoots and dug camas bulbs from freshly laid Quaternary soil. Like the first trees, the forebears of the Salish, Kootenai, and Pend d'Oreille tribes took root in western Montana as its modern landscape took shape. The Salish were the southernmost of these tribes and the people most closely tied to the Mission and Bitterroot valleys.

They wandered and lived between mountains, journeying east for bison and west for salmon that swam upstream from the distant Pacific. They watched towering old predators disappear like falling stars. *Arctodus* was gone, then the dire wolf disappeared. *Smilodon*, the saber-toothed cat, starved for lack of outsize prey. Soon, the people were alone with grizzlies, moose, and a few other species in their recollection of the former world.

Their stories echo with old beasts. There was an elk monster living down in the Bitterroot Valley until Coyote killed it. There was a ten-mile serpent in the Jocko Valley, just south and west of the Mission Range, with its mouth at the crest of Evaro Hill and its stomach near the town of Arlee. Coyote killed that creature, too.

Some things disappeared from their world. Others arrived.

Neighboring tribes including the Blackfeet infringed on Salish territory, pushing westward with newly acquired guns in their hands. A wave of illness followed on their heels, more difficult to withstand than Pleistocene winter.

If Salish history were a drumroll, with each quick beat representing a generation between the end of the last ice age and the start of the twentieth century, these are the last several taps:

A beat, and the first traders came overland from the east. Another, and a Catholic mission stood in the Bitterroot.

A beat, and some of the chiefs made their marks on the pages of the Hellgate treaty, ceding 19,000,000 acres to the United States Government and reserving 1,245,000 for the tribe. Another, and the mission moved to St. Ignatius. Forced on by a company of army regulars, the Salish people marched north from the Bitterroot.

A beat, and most surviving Salish shared a corner of their former lands with what remained of the Kootenai and Pend d'Oreille tribes. On the Flathead Reservation, they were forced to try their hands at farming. Another, and hungry white men chewed the reservation's fringes, lusting after water and fertile soil.

The Salish bore witness to the birth and expansion of their cultural and geographic world, then to its fragmentation. The Dawes Act was passed by the United States Congress in 1887. Thereafter, in the early part of the twentieth century, the tribes endured the allotment of the Flathead Reservation, in which the head of each tribal household received title to 160 acres, other adults received 80 acres, and the balance of the reservation was opened to nontribal homesteading and purchase. A rapid liquidation of farmland followed. Before halting the allotment process, the tribes lost more than a million acres that had been guaranteed to them in perpetuity.

That land, initially claimed by farmers, ranchers, and specu-
lators, has passed to outsiders of every stripe. Today, the reservation boundary includes Amish colonies, the fiefdoms of absentee
millionaires, survivalist hideaways, old hippie communes, organic
market gardens, and a Tibetan Buddhist monastery. A large por-
tion of the fertile soil remains in agriculture, though that way of life
is threatened everywhere by growth and new construction.

The reservation's demographic patchwork is overlain by the
usual tensions of the American West. Water wars — the largest lasting more than a decade and pitting the tribes' treaty rights against
the claims of irrigators — smolder in court and flare on ditch banks
in late summer. Irreconcilable neighbors squabble over grazing
leases and rights-of-way. Noxious weeds make inroads and must
be destroyed. Realtors converge on failing ranches like magpies on
a carcass, ready to pick apart and sell the dream.

The reservation is home to people who love grizzly bears and
wolves, and others who would rather see predators wiped out. It
holds happy families, paroled criminals, snowbirds, and rednecks.
Everyone agrees that the mountains are beautiful. It is perhaps the
only point on which they see eye to eye. That, and the fact that
as the Mission Valley grows crowded — as its expanse, plentitude,
and access to wilderness are fractionated — they all feel a creeping
sense of loss.

I DROVE onto the reservation on a spring morning in 2016,
meaning to meet Stacy Courville in person for the first time and
figure out whether People and Carnivores might be able to work
with the tribes on issues related to grizzly bears and wolves. Cheatgrass was greening in the borrow pits beside Highway 93, the sky
was clear, and the soil was dark with snowmelt.

Heading north from Missoula, I climbed Evaro Hill, crossed the reservation boundary, came down through the town of Arlee, and followed the serpentine route of the Jocko River. I tried to sound out the Salish place names on roadside signs, got thwarted, and settled for reading the English translations — Spring Creek and Coming Back Down to the Water's Edge.

After Ravalli, I flogged my truck uphill to a place where I could see the Mission Valley opening northward. Stopping there, I waited for the sun to top the peaks. When it did, and yellow rays spilled over the mountains, I could see the valley floor. Gravel roads crossed it at regular intervals, as if a wide-meshed net had been thrown over all the arable land. Cars and trucks traced the warp and weft, their headlights burning. Houses, barns, and outbuildings dotted the panorama, some of them sending up pale smoke. Rising clear of the peaks, the sun flashed on water running in the valley's many irrigation ditches.

The landscape cannot be called unspoiled or empty. As of the 2010 census, the Flathead Reservation was home to 28,359 people, with the majority of them living in the Mission Valley. Neither can the reservation's demographic mix be described as primarily indigenous. Though the Dawes Act was repealed nearly a century ago, the privatization of tribal land has left an enduring mark: Most of the Flathead Reservation's residents are white.

According to the Montana Governor's Office of Indian Affairs, the Confederated Salish and Kootenai Tribes have 7,753 enrolled members, with approximately 5,000 members living on the reservation. Non-members, then, outnumber Salish, Kootenai, and Pend d'Oreille people on the reservation by more than five to one. Seen in this light, the houses, barns, and canals covering the valley floor are as much a record of cultural displacement as ecological change.

The tribes prove their immense resilience against this backdrop, struggling mightily to maintain their identity, assert treaty rights, and recover lost land. Against long odds, they preserve a measure of what they have always been.

Damage and hope are evident all over the reservation: the former boiling up in a rash of suicides or the word "Rapist!" scrawled across a prefabricated house; the latter manifest in a town's fervent support for its high school basketball team. Those young men, the Arlee Warriors, won the Class C state championship two years in a row, both times beating a private, well-to-do Christian school. Their story ran on the front page of the *New York Times Magazine*. A motorcade of cop cars, ambulances, and fire engines met the Warriors at the reservation's edge, bringing them home on a rising wave of noise. A siren is a fitting sound, keening and singing, as closely bound to loss as to aid.

CONSCIOUS OF the place's long and unjust story, I put the truck in gear and drove into the valley. Roadside fields looked fertile and serene. Passing cars were full of commuters headed for Missoula. To my east, the mountains formed a wall as foreboding and wild as any I have seen, and I followed it north toward the town of Polson.

Driving toward my meeting with Stacy, I knew enough about the tribes' approach to wilderness and wild animals to be hopeful. Throughout the twentieth century, after seeing their valley opened to settlement and broken out to cropland, foresighted tribal members worked to safeguard the high, steep country on the reservation's eastern edge. Their efforts bore legal fruit in 1975 with the passage of tribal Ordinance 79A, a document protecting 91,778 acres of land as the Mission Mountains Tribal Wilderness area — a five-mile-wide swath of peaks and cirque lakes running the length of the valley. Borrowing language from the 1964 federal Wilderness Act, the ordinance prohibited road building, logging, and permanent inhabitation in the Mission Range.

We can all thank them for it, because the Missions are important mountains. East of the knife-edged ridges at the reservation's edge, wild country spreads north into the Canadian Rockies through the Bob Marshall Wilderness and Glacier National Park.

This makes the tribal wilderness area the far southwestern tip of an enormous untrammeled ecosystem. A wolverine could feel a restless itch in the Yukon, walk south for weeks through forests and over mountain ranges, and come to stand on McDonald Peak looking over the Mission Valley. Just below her, nestled against the foothills, would be Millie's Woods and the field where Greg Schock grows corn. In all her traveling, the creature would have crossed only one major freeway, a pair of two-lane highways, and a handful of paved and gravel roads.

With so much adjacent wilderness, it is no wonder that the Missions still contain every fish, bird, plant, and mammal that met Lewis and Clark on their trudge across the West. But for all that diversity, anyone who has walked the western edge of the Mission Range knows that the mountains belong to grizzly bears. A certain feeling comes to hikers up there: a tight thrumming in the stomach, a tendency to startle, a consuming interest in shadows.

The gut does right to twist in these mountains. The eyes, flitting to the low shape of a burned stump, are wise. Science has corroborated what the body knows: A very high density of grizzlies exists in the Missions, particularly near McDonald Peak, northeast of St. Ignatius.

Since 1982, the tribes have closed ten thousand acres around the peak to all forms of human travel, commerce, and recreation every year between July 15 and October 1 to minimize the disturbance of bears and the danger to humans. Nobody enters, and for two and a half months, bears move unseen through the drainages above Post Creek and Schock's dairy. They drink from the Ashley Lakes and prowl groves of soft-needled larch. Nobody is there to see or bother them. For a little while each year, so long as they keep to the mountains, the grizzlies live unmolested in a vestige of the older world.

WALKING INTO the corrugated-steel building that houses the Natural Resources Department of the Confederated Salish and Kootenai Tribes, the first thing that I noticed was a fish tank built into a wall. Lazy, blunt-jawed rainbows and cutthroats circled, looking through glass at deer skulls mounted on the far side of the hall. A bearskin — splayed and with its mouth wide open — hung on the wall beside the stairs. Climbing to the second floor, I resisted an impulse to sink my fingers into its cinnamon fur.

Stacy was waiting in an office wallpapered with maps and informational posters, behind a desk littered with printouts and telemetry equipment. Sturdy and goateed, with close-cropped salt-and-pepper hair hidden by a ball cap, he looked up from working to shake my hand.

After walking down a hall, we took chairs at a table in a small meeting room with shelves crammed with books and bound reports.

"Shannon Clairmont," Stacy said when a slight, darker-skinned man joined us. "We work together on forest carnivores — wolves, bears, furbearers —"

"Pretty much anything that causes trouble on the south half of the reservation," Shannon added, reaching across the table to shake my hand.

"Right," Stacy said. "And Shannon, this is Bryce. He's working for — who is it again?"

"People and Carnivores," I said, producing from my notebook

a newly printed business card with my name on one side and the group's logo — a grizzly track overlain on a thumbprint — on the reverse. I felt clumsy handing the card across the table, foolish while the men examined it.

I told them I had been hired by the group to expand their work in western Montana. Confessing that I was by no stretch a biologist, I talked about my experiences running ranches on the edges of Yellowstone and my desire to work on issues related to grizzly bears and wolves.

Leaning back in his chair, Stacy watched me closely.

"Issues?" he said, crossing his arms and turning toward his partner. "What do you think, Shan — we got any issues?"

Shannon cracked a grin. "Heh," he said. "Might be a couple of those."

Stacy launched into a description of the reservation's wolf packs, noting the home range and history of each, and the number of times they had been "removed" because of their depredations on livestock. Soon, he was warming to a discussion of bears, with Shannon breaking in periodically with details.

Listening to them, I was struck by the difference between the two men. Stacy's bearing is ponderous, even grave. Shannon is freer in his movements and readier with his smile. Though he let Stacy do most of the talking that day, Shannon never stopped interjecting in quick, bright bursts. In contrast, Stacy chose his words deliberately, laying them down as a mason sets bricks, examining each one for faults.

Sometimes he let seconds pass between one thought and another while I struggled against finishing his sentence.

"The Mission," he said in his methodical way, "has always been grizzly country. It was when I started this job in '95. When I was growing up, too, in St. Ignatius. There were grizzlies then, but we didn't see them in the valley like we do now.

"First ten years I had this job, we hardly trapped any grizzly bears. In 2005, we started collaring for research. We still trap for research, but it's the conflict calls that have gone up."

"The hardest part," Shannon added, "is getting people to change their ways, particularly new people moving into the valley. They bring in bear attractants — trash, pet foods, livestock feed."

"Hobby farms," Stacy said with a critical shake of his head, and he began to describe the most recent of the Mission's many agricultural and social changes: an influx of small-scale farmers and back-to-the-land escapists that had begun with the turn of the millennium and increased in subsequent years.

I knew something about this latest wave of immigrants. Living in Missoula, I was conscious of a steady northward trickle of men and women in their twenties and thirties. Finding that land on the reservation was fertile and comparatively cheap, these would-be farmers left the city to grow row crops or raise meat in the Mission Valley. They bought property when they could and leased it if they had to, and their produce fed Missoula's burgeoning local food scene.

One particular place comes to mind, a spread with a disorganized charm. It cannot properly be called a hobby farm because the owners work it for their living, but it is smallish, organic, and cast in a new mold. The market garden, planted with a dozen types of vegetable, becomes a quilt in summer. Pastures are divided with temporary electric fence so animals can be rotated through them to avoid overgrazing. Chickens run amok. Hogs turn the sod with evident joy. When I was there, I did not want to leave.

That farm looks different from the cow-calf, hay, and grain operations that had prevailed in the valley during the last century, and it functions on a different scale. The old guard shipped off the fruits of their labors by the semitruck or train-car load and spoke a language of hundredweights and commodity prices. The new farmers haul smaller quantities of meat and vegetables to Missoula each week, to restaurants and open-air markets. "Local," "heirloom," and "organic" are their sacred words.

Some tension simmers between farmers of the old and new sort. It could hardly be otherwise when the two groups so often differ over politics, the use of irrigation water, the ethics of growing genetically modified crops, and how to control the weeds that spring from the Mission's deep soil.

Stacy told me that he found it hard to connect and communicate with the newcomers because they moved in different social circles than the old-guard farmers. It was a struggle to teach them how to share the landscape with large predators.

The main problem, Shannon said, was that the new people kept small animals. "Chickens, goats, pigs, llamas. You put those near the wilderness or close to cover where a bear can travel, like a riparian area, and you're going to have trouble. All those little animals are an easy food source for a grizzly. If they're not kept behind an electric fence or put away, bears get them."

Grizzlies, Stacy told me, were particularly fond of poultry. Responding to chicken raids had become a regular and irksome part of his and Shannon's summer work.

"All they leave are gizzards and feet," he said, and with a grim shake of his head, he explained the dire consequences of such a feast.

Because a chicken-eating bear — or a sheep eater or a grain-bin raider — was unlikely to stop of its own accord, Stacy and Shannon had to trap and relocate it. That difficult, dangerous work yielded only mixed results as bears tend to find their way back to familiar country.

When grizzlies could not be broken of their bad habits and continually ended up in trouble, Stacy and Shannon were responsible for killing them. Neither of the men relished that work, and Stacy was keenly frustrated by the idea of shooting an endangered animal because it couldn't resist eating a few of the nation's most common birds.

"Getting into chickens," he said, "is a death sentence for a bear. It's avoidable, too, if people are willing to put up some electric fence. Folks who live here need to get that through their heads, and they need to understand how well grizzlies move around the valley."

Setting his hand palm down on the table, he rubbed it in a slow circle.

"Grizzlies own this valley at night. They go all over for food — walk the roads, even. And for as many bears as we have, there are *very* few conflicts. So the bears are good at living with us. But these new people, they're no good at living with bears."

"We educate people as much as we can," Shannon said, "with bear-aware posters and flyers, and articles in the paper. I don't know how many of them take it seriously until something goes wrong."

The men paid frequent visits to farms and homesteads that seemed ripe for disaster. They dispensed advice and warnings and stuffed mailboxes with information. That was the most they could do on private land. Because so much of the reservation has been sold outside the tribe, they could not legally compel a farmer to bring her sheep in at night or electrify a chicken coop.

Stacy described the challenge of convincing people to take action, how slow the pace of change could be until something went catastrophically wrong. Struggling against reluctance was half the battle, he said. The other half was trying to keep grizzlies up higher in the mountains.

The bears had natural food sources up there. For as long as

whitebark pines have grown along the tree line, squirrels have been gathering and caching oily, nutritious pine nuts in preparation for winter. And for just as long, grizzlies have been smelling out and devouring those hidden stores. Grizzly bears have other alpine resources, too. Vast swarms of insects, principally army cutworm moths, gather each fall beneath glacial boulders on slopes near McDonald Peak. Flipping these stones, bears can uncover enough calories to last through winter.

"But none of that can compete with what's in the valley now. We've had apples for a long time and livestock. Now, more and more, bears are learning to feed on field crops like corn.

"On the edge of the Mission Mountain Front," Stacy continued, "is a place we call Millie's Woods, which has always been high-quality grizzly habitat. I've known about it since I was a teenager, pheasant hunting. There's a dairy just a half mile south of those woods, and the farmer, Greg Schock, grows corn and chops it into silage to feed his dairy cows.

"In 2002, Greg seeded a little field at his parents' place, and bears found it. They didn't take much at first. But year by year the problem was growing, and Greg started complaining about the damage and what it cost. He was right to talk — he was feeding a lot of bears from August through when the corn was cut in September."

From the start, Stacy and other biologists saw the peril of bears learning to eat corn in the Mission Valley. In 2005, the tribes' Natural Resources Department partnered with a conservation group to build a grizzly-proof fence around Greg's field. The expensive, seven-strand, electrified barrier worked well at first, although its lowest wire, just a few inches off the ground, was in constant danger of being shorted out by thick growth. Stacy, Shannon, and other tribal employees fought the grass with string trimmers for a handful of years until the crop exhausted its soil and Greg decided to move his corn south to a larger hundred-acre field. Nothing had been done to protect the crop there, Stacy said. With a perimeter of around 11,000 feet, a fence would be too expensive to build and too difficult to maintain.

I asked him if bears still came to feed on the corn.

"More every year, and they do a lot of damage. Up to ten thousand dollars' worth in a season, Greg says, and sometimes he has to buy grain to get his cows through winter."

The situation was more than a hazard to Greg Schock's livelihood. As a limitless source of food, the corn concentrated a large number of bears in a small area. Up to ten grizzlies had been observed leaving the corn in a single morning, and about as many bears were believed to remain in the field day and night while the ears were ripe. To have so many grizzlies converging in a landscape full of farms and homes was a recipe for conflicts and violent encounters.

"This is a learned behavior," Stacy said. "For years we've been producing grizzly bears that know how to eat corn. The females stick around, but males disperse to other places. No matter where they go, they know corn. When they find it, they stay and feed until it's gone.

"You can watch it spread: When new bears come into the field, their scats will be full of apples or chokecherries. But after they try corn, they never quit.

"We've got to do something about it," he continued after a pause. "Because with more people in the valley every year and grizzlies expanding their range, the problem isn't going anywhere."

A tinny melody pealed through the room, cutting him short.

"Look for the bare necessities,
The simple bare necessities."

Fishing a phone from his pocket, Stacy studied the screen and

turned it to Shannon. He silenced the call on the song's second refrain, and the three of us sat quietly. Shannon stretched in his chair
and looked once over his shoulder toward the door. Feeling that I
was keeping them from important business, I reiterated my desire
to work together, took leave of the men, and headed back down the
valley.

DRIVING, I tried to imagine the shaggy bulk of a bear moving
through a cornfield. The image eluded me. Corn brought Kansas
to mind while grizzlies conjured Glacier and Yellowstone. Though
I couldn't picture a bruin biting into an ear, the consequences of
bears getting hooked on corn in the Mission Valley were clear
enough. Passing south through the small towns of Pablo and
Ronan, I thought about how that habit might impact the species'
future and their chances of thriving in western Montana.

When I first went to work for People and Carnivores, I learned
that one of the group's great aims was to reconnect the grizzly bears
living in and around Yellowstone National Park — an area known as
the Greater Yellowstone Ecosystem, or GYE — with their more numerous cousins in the Northern Continental Divide, an ecosystem
stretching northward toward Canada from the Mission, Swan, and
Blackfoot valleys.

Yellowstone bears are, in a genetic sense, marooned. Though
they number more than seven hundred and have expanded their
range outward from their core habitat in the park, Yellowstone
grizzlies do not regularly encounter and mate with the thousand
or so bears making their home in the Northern Continental Divide
Ecosystem — the NCDE.

This isolation leaves Yellowstone bruins vulnerable to the ebbs
and flows of ecology and reproduction. Without connections to

larger populations and more expansive habitats, a series of bad years or an outbreak of disease could be ruinous for GYE grizzlies.

Decades ago, a map of Montana's grizzly distribution looked somewhat like a thick exclamation mark, with GYE bears forming the dot and NCDE bears living in a line stretching down from Canada along the spine of the Rocky Mountains. Mission Valley bears live at the southwestern terminus of that line.

Under the protection of the Endangered Species Act of 1973, grizzlies have become more numerous in both the GYE and the NCDE. Dispersing in search of mates and unoccupied territories, bears are entering areas where they have not been seen in half a century. Some have tried their luck—with grim results—far from the mountains on the cropped prairies of north-central Montana. Others are moving north and south from their core habitats along more promising lines in a manner that may yet bridge the gulf between Yellowstone and the Northern Continental Divide.

Still, reconnecting those two populations is a complex and uncertain business. Grizzlies are difficult, inscrutable creatures, but this much is sure: The Mission Valley, as the southwestern extent of a grizzly population running northward to the Arctic, is a likely source of dispersing bears. It is a jumping-off place from which grizzlies may recolonize mountain ranges and valleys that once were theirs.

And there is something else: Straight southwest from where I was driving home toward Missoula on Highway 93, the timbered hills on the far side of Interstate 90 hump up to mountains that run uninterrupted to the Selway-Bitterroot Wilderness. South of that is Idaho's Frank Church–River of No Return Wilderness, the largest roadless area in the continental United States. Together, the two blocks of land comprise a five-million-acre expanse that harbored many grizzlies in the past but has been empty of them since the middle of the last century. The last confirmed kill of a grizzly bear down there was in 1932, and the last documented tracks were seen in 1946.

Nobody knows where a Mission Valley bear might go when it takes a notion to travel. A yearling might walk southwest into the vastness of the Selway-Bitterroot in search of vacant habitat or strike southeast toward its Yellowstone cousins. Wherever it goes, that dispersing boar or sow will carry the habits and knowledge of its youth.

What bears are taught in the Mission Valley is crucial. If they start raiding chicken coops, they will almost certainly be shot for doing so in the years that follow. If they pick up the habit of feeding in cornfields, they will continue to do so in other places and thereby come into conflict with farmers. If they learn to keep to the high country instead, steering clear of humans and their crops, grizzlies are more likely to prosper and survive.

SEEN IN this light, Schock's cornfield was a gateway to ruin. No better situation could have been devised for teaching grizzlies to feed on agricultural crops, and no worse place could have been found for it than on the southwestern edge of the Northern Continental Divide. As I crested Evaro Hill and sped south toward Missoula, it struck me that addressing the issue at Schock's field might improve the behavior and survival rates of dispersing bears, contributing to the restoration of grizzlies to parts of their historic range.

Coming into town, I wondered how I might tackle the problem. After a decade of ranching, I understood the predicament well: Encircling a hundred-acre field like Schock's with shoulder-high, seven-wire electric fence would cost fifty or sixty thousand dollars. People and Carnivores didn't have that kind of money. Neither did the tribes, and I couldn't imagine Schock's family-run dairy footing the bill on their own.

Still, I wanted to do it. I wanted to fence Schock's field from the moment I heard about it — partially because I thought it would do some good for grizzly bears and partially because I lacked for physical labor.

In retrospect, I think that I would have taken on any hard job that crossed my path. Up until then, my work for People and Carnivores had required a great deal of talk and little action, and this had left my hands itching for a hammer to swing. Spring had come, and just as a bear knows that season as the moment to rise and seek winterkill, I recognize it as the time for bending to work and hardening my muscles under the warming sun.

Perhaps because I was so hungry for exertion, an idea came to mind. Weeks before, James Jonkel, a bear biologist with Montana Fish, Wildlife & Parks and a respected figure in the world of large-carnivore management, had told me about a study underway in the nearby Blackfoot Valley. Up there, two valleys east of the Mission, a Montana State University graduate student named Brittani Johnson had spent a summer building small electrified enclosures in areas frequented by grizzlies. Leaving road-killed deer as bait within and setting up motion-sensitive cameras at each site, she had collected video recordings of bears interacting with electric fence.

Brittani aimed to learn whether a particular design of three-wire electric fence constituted a significant impediment to grizzlies moving across agricultural landscapes. It was an important point to settle, as the fence in question was in wide use throughout Montana. I had built miles of it over the years and had always felt good about doing so because of how kindly the design treated ungulates. Deer and elk jump right over a three-wire fence. Antelope slip under, hardly breaking their stride. When I replaced a barbwire fence with a three-wire, high-tensile one, I could rest easy knowing that I would seldom have to cut a dead or entangled animal from the wires.

Bears approach fences differently. I have watched sows and cubs move through barbwire as if it wasn't there — they are that lithe and that tough — but electric fences stop them in their tracks. Perhaps it is because they lead with their noses or because soft, wide paws put them in excellent contact with the ground. For whatever reason, a bear feels and respects a shock. Brittani's study confirmed this and revealed a striking trend: In a full research season, no grizzly had yet managed to cross a three-wire fence with strands hung nine, sixteen, and twenty-five inches off the ground.

If a barrier like that could deter bears on a small scale, I reasoned, why wouldn't it work around Schock's corn? A low three-wire fence could be built much more cheaply than a traditional bear fence. Every aspect of constructing it would be easier. The line posts could be smaller, and the braces that carried the strain of the wires could be widely spaced. I did some rough math. Components for such a fence would cost around forty cents per foot — less than a third of the materials cost of a traditional fence. And because of the lighter components, the work could be done without hiring a contractor, which would further cut the project's expense.

Putting a three-wire fence around the corn seemed like a good, simple solution. I resolved to draw up a budget, pitch the idea to Lisa at People and Carnivores, and get in touch with Greg Schock. I pulled into home wondering how soon the bears would come down to the valley floor and when I would need to start building fence if I meant to keep them out of the cornfield.

MILLIE GRAZED among the sheer and thickly treed peaks above the Mission Valley, cropping new grass on a south-facing slope and turning over stones in search of anthills. Lifting her nose, she pulled a draft of the burgeoning, fecund world and parsed it carefully: aspen bark and moldering leaves; songbirds in the trees; her own reek, blunt and pungent after months in a hole; damp earth; the coniferous tang of mountains; water.

She smelled through all this, shuffling scents until she found a ripe whiff that mattered. Millie tossed her head, teasing information from the mysteries of the air.

Once she had it, she walked up the line of the wind. The cubs were with her, following closely. Millie was a fine mother. She went ahead, stopped, turned, and waited for them to come on. Strong with milk, the two small female cubs labored over the mountainside, pressing on with ursine stoicism.

Months before, as winter settled onto those same sheer, thickly treed peaks, a bull elk had exhausted himself in rutting, sustaining small wounds and going too long without feeding. After the lusting season, he went to shelter in the timber. The north wind blew across miles of alpine ice, ever colder, and found him sleeping alone among pines. Hushing in the branches, the wind put its finger on the bull's nose. Experimentally, it lifted the long hollow fibers of his winter hair. That greatcoat would have been equal to any weather if fat had remained beneath it, but the elk was poor and gaunt.

I have seen many things in the mountains but never a bull elk dying of cold. I cannot say if such an animal shivers or moans. The business is probably slow and quiet, and the antlered head settles to rest in the snow. From hunting, I know that it is a long time before heat escapes the body, and a dead elk below zero will be days in freezing through. A lot of heat is in a bull — even a weak one — much more than is in a human being. An elk is a musky, living furnace. The smell of a herd, even when it is a day old, burns in the lungs. In death, that heat and smell are long in going.

If a person must bet on anything in Montana, choose the cold. Take odds on it against the burning vigor of elk or well-laid plans. Bet on cold to win in the end, to break the pipes and grip all creatures.

Winter enfolded the bull. Curled tightly, swathed in drifts, he waited in the dark for spring. When that season came, light burned through the trees and the snow ran water. The elk's coat turned slick and matted, looking wretched when his uphill shoulder melted free.

Black-eyed ravens spied him through the canopy. They struggled with the hide, opening small holes. Flies set their blind maggots to growing.

The stink of rot went out with the smell of meat, unfurling ribbons through the grove and over the mountain's thawing face. They rode the wind, thinner but persisting.

A bear can smell very well — much better than a bloodhound. Grizzlies are capable of scenting carcasses at a great distance and have been observed walking upwind for miles toward a meal. Their olfactory world is a tapestry without edges, with hungry boars and sows following individual threads.

MILLIE'S NOSE missed nothing. She crossed steep open faces below the peaks, interrogating every breath. The breeze drew her into a ravine through a swollen stream. The cubs forded hesitantly and shook like dogs on the far shore.

Sure of her bearing, Millie ascended toward a stand of trees. She lengthened her steps until the cubs struggled to keep pace. Unceasing, she pushed through close-grown timber toward a melting drift patterned with corvid tracks and bird shit. Millie neared it warily, looking in all directions through the timber, huffing at her cubs until they crowded tightly to her flank. She waited, watched, and listened, covering the last distance only when she felt sure she was alone.

Sweeping her big front paws, she gathered the snow inward toward the keel of her chest and flung it back along the length of her body. Working downward, she disinterred the elk's front leg and shoulder. She freed the chest and stomach next, and when her long claws caught on skin, she slit it as if breaking an envelope's seal.

Viscera spilled out — lungs, dark liver, and the veined rumen — and Millie consumed them with ravenous efficiency. She bolted down organ meat, frozen or thawed, and when it was gone, she clamped her teeth around the bull's upper foreleg. Pulling with the mad strength of hunger, laying back against the strain like a dog in a tug-of-war, she jerked the winter-killed bull from the snow. Falling upon the carcass, she peeled the hide and bit through icy muscle.

Tearing was her modus operandi, and the cubs watched closely. Seeing the carcass fall to pieces, they dodged in and chewed scraps with their small white teeth. Though they bumbled and dithered, Millie let them learn to eat. She ripped, chewed, and swallowed until the edge of her winter hunger was blunted. She gorged a good while longer until she could get no more down, then she walked through the trees to where afternoon sun patterned the snow.

Millie kept still, her belly swollen with the year's first meat. She let the needle-mouthed cubs suckle as the light turned from yellow to blue. Darkness, when it came, did not chill her.

I HAD a plan in mind when I went to meet Greg Schock for the first time. Talking with Brittani Johnson about the latest results from her fence study in the Blackfoot Valley, I had learned that the low, three-wire design was still standing up well to pressure from grizzlies. Deciding that it would be worthwhile to build the same sort of fence around Schock's cornfield, I had done my best to convince Steve and Lisa at People and Carnivores of the project's merits.

The broad strokes of my argument were as follows: First, that becoming dependent on agricultural crops was a big problem for bears in the Mission Valley. Feeding on corn brought grizzlies into close proximity with people, increasing the chance of dangerous encounters. Second, that the problem's scale would increase in the future as climate change, genetically modified seeds, and improvements in irrigation technology made it possible and profitable to grow corn commercially in high, cold places like western Montana.

Plowed fields across the valley testified to the issue's scope. The reservation's rising grizzly numbers suggested that the future would see more bears dispersing from the mountain and encountering corn.

It was a large long-term problem, I told Steve and Lisa, and one requiring an economically viable and scalable solution. The three-wire fence could deliver on that front. It was cheap enough to build around a ninety-acre field, and its cost could be offset within a couple of years by the value of corn no longer eaten by bears. Best of all, the fence posed no barrier to deer, elk, or most other wildlife. Even a fawn could clear a top wire that was twenty-five inches high.

I took comfort in the fact that a field fenced that way would not stop most wild creatures from moving freely across the landscape.

Enclosing Schock's field would be an experiment, I admitted, because short three-wire fences had not been used to exclude bears on a large scale before. But if the experiment worked, it would prove out a powerful new tool for coexisting with grizzly bears.

Lisa and Steve liked the idea well enough to approve the project, provided that I could get the landowner on board. We found partners, too. Stacy and the tribes volunteered their help, and the United States Fish and Wildlife Service contributed funding for materials through their Partners for Wildlife program. With all the pieces in place except for the one that mattered most, I called Greg Schock and set up a time to meet.

RIDING HIGH on the expectation of building something new and useful, I drove to Schock's place, shook his callused hand, and was shortly riding shotgun in his truck on a tour of the dairy.

"How it started with the grizzlies?" he said. "How it started was I called Stacy and told him there were bears in the corn, knocking it down. He said that bears didn't eat corn, and it was probably raccoons. Told me raccoons can do a lot of damage to a cornfield — more than you'd think, for how small they are. But I said he had to see it, so he came out and we went down to the field where I showed him a pile of — well — I showed him a pile. 'Take a pretty big raccoon to make that,' I said."

Leaning back in the driver's seat, Greg Schock eased the pickup along East Post Creek Road. The truck was a one-ton dually, but even so, Greg looked as though he had been wedged into the cab. He is a large man, not overweight or extraordinarily tall but substantial in every limb and aspect.

Thick-fingered hands rested on the steering wheel, and a thatch of graying hair fell nearly to his brow, ending in a straight-cropped line. From the far side of the cab, with the sun slanting through the window, the chiseled heaviness of Greg's features made him look as though he had been carved roughly from stone. A lifetime of working on and beside diesel engines had taken some measure of his hearing, leaving him with a voice just shy of a shout. Talkative from the start, Greg thundered affably about the achievements of his children, the particularities of dairying, and his difficulties with the grizzlies.

"We planted the first corn in '02, at Dad's place. That was the year I brought Stacy out and showed him the pile. Bears didn't bother the corn much that year — worked the edges mostly, came and went so that we hardly saw them."

We reached his dairy barn, a faded, yellowish Quonset hut in which seventy or eighty Holsteins lay in rubber-floored stalls or loafed on grooved concrete.

"The thing about those stalls," Greg said as he idled to a stop, "is they're shaped so one cow fits into each, and when she lies down, her hind end hangs off the concrete lip. That way the manure falls into the center aisle, where we can scoop it up with the loader. They're slanted, too, so the cows always lie the right way, keeping their heads up."

As if on cue, a cow shat a thick green pie.

"Do they get out on pasture?"

"They do. Sometimes they go out to the yard there." He pointed to a close-cropped field. "And they walk every day from here to the milk house.

"Course, they're out on pasture when they're dry. But this isn't like running beef cattle. This isn't what you're used to. No, a dairy cow works hard. She's got to make sixty pounds of milk a day to be pulling her weight. Got to do that every day, and she can't do it on pasture. So she needs corn and plenty of alfalfa. Even so, four years, or five, maybe, and a dairy cow breaks down. Look at it that way, and you can see how there's always going to be something here for a bear to eat even if we didn't have the corn."

A cow stood off by herself, reaching back in a straining U to lick at her flank. Greg watched the animal closely, and I watched him, thinking that he seemed amiable enough, that he showed none of the antipathy toward large carnivores that is so common among ranchers. I supposed him to be a peaceful man—a live-and-let-live sort.

His dairy, though, was messy. Derelict irrigation pipe snaked haphazardly through some of the fields. Large piles of livestock bedding and manure were intermingled with old machines, and the fences were in a state of disrepair for which I'd have lost every ranch job I ever held. The pastures nearest to the dairy barn were overgrazed. With the grass chewed down beyond the point of resilience and regrowth, weeds had sprung up and flourished.

In spite of the disarray, I did not take Greg for a lazy man. Tough, square hands and bowed shoulders testified to a life of hard work.

He studied the cow, palms curling over the steering wheel, face expressionless and slack. For a moment, his bearing matched the state of the land, which had been grazed, watered, and cropped to exhaustion.

"You know beef," he said, stirring. "You're young, but even so, I guess you've heard how it used to be. A guy could haul a stock-trailer load of calves to auction with an old pickup—bring twelve, fourteen head—and drive home from town in a brand-new truck with what he made. Now you want to buy that new truck, it'd mean selling sixty, seventy head—a semi load. Well, the dairy business is like that, too. Costs go up all the time—fuel, machinery, what grain I need for the ration and can't grow here. And milk, well, you can go down to the grocery store and see how much its price has changed."

We headed north on Hillside Road, then east toward the mountains on a spur.

"Thing is," he said, "I like bears. I like watching them and I know they belong here. They don't bother my cows, and I don't mind them cleaning up carcasses. It's just that I can't afford to keep losing so much corn. With the market like it is and margins like they are, I can't feed them."

He stopped beside a tall, weathered, seven-wire electric fence to show me where he had planted the first corn and where the crop had grown until, needing more, he moved it to a bigger field near his house.

"Stacy might not have told you," he said, breaking into a smile that creased his face deeply enough to hide his eyes. "But the first year we had this fence—2005—a bear got in. Not sure how he did it—maybe a gate got left open or grass shorted out the bottom wire—but there was corn knocked down and we started seeing piles again."

"How'd you get it out?"

"Never could. Stacy and Shannon came down, and they brought the government trapper with his bear dog. But that dog—soon as they let him out of the kennel, he went to the fence and peed on it. Well, the fence was hot. It sent that dog five feet in the air, and for the rest of the day, he stuck to the trapper's side. Couldn't get him to run through the corn for anything. Stacy and Shannon spent a couple hours driving up and down the rows shooting cracker shells and honking. I watched the whole thing from my pickup on the road, but the bear never showed. He just hunkered down in there, and nobody was going to come in after him.

"That's the thing," he said with a nod of his craggy head. "They get in there, you can't ever get them out."

I wondered what it would be like to try to chase a grizzly from a stand of corn, picturing the booming racket of noisemaker shotgun shells, the way a big animal would set the stalks in motion.

"We talked a little on the phone," Greg said. "But tell me again what you've got in mind. What is it you want to do?"

I told him that after learning about the situation from Stacy and Shannon — how many bears got into the corn and the damage they did — I had talked with the conservation group that employed me. The executive director, I told him, agreed that keeping bears out of his corn was a project worth working on.

"That is," I said, "if you want to try to do something about it."

"Which I do," Greg said. "But I don't have money to put up. With the price of milk —"

"You wouldn't have to spend money. Not with what I've got in mind. After I got the go-ahead from my boss, I got in touch with a woman who's studying grizzly bears in the Blackfoot for Montana Fish, Wildlife & Parks. She's been testing a new kind of electric fence up there. It's just over two feet tall, and you wouldn't think it would work, but she's been testing it on grizzly bears. She's been setting up exclosures on the edge of the Bob Marshall Wilderness, putting carcasses inside of them. So far, the fence is turning bears."

Greg looked through his open window at the field. Behind the defunct electric fence, alfalfa seedlings were spreading green across the hardpan soil.

"Two-foot tall," he said. "And it works?"

"Seems to, but I'm with you: It's hard to believe. That's why this thing would be an experiment. We would—I would—build an electric fence around your field like the ones in the Blackfoot study. Then we'd watch what happens to see how the fence works. And if it does work, that would be big news, because a three-wire fence is a lot cheaper than what you've got there." I gestured at the over-grown wires around the alfalfa field. "And it might be a way to keep bears out of crops all over the state."

Greg considered this.

"You'd build it yourself?"

I told him that I would.

"And you'd buy the materials?"

I explained that the United States Fish and Wildlife Service would foot most of the bill for supplies, with People and Carni-vores covering the balance.

"Who'll maintain it?"

"For the first year, and as long as we're running tests on it, I'll take care of the fence. After that, it'll be on you."

Craning his thick neck toward the peaks, Greg paused.

"One thing I worry about," he said, "and maybe I shouldn't, is where will all those bears go if we fence them out of my corn?"

"Back to the mountains, I hope."

"Maybe. Maybe, but I don't think so. They've been coming down after apples for a hundred years. And look at the valley now." He gestured expansively at the barns, houses, and tidy fields below us. "All the people who have come in. New to the valley, and every one of them wants a garden, chickens, a bunch of sheep. How's a bear supposed to keep out of trouble down there?"

He coughed into a massive hand and was silent.

"Have at it," he said after a while. "No matter what, we'll be bet-ter off than we were before."

• • •

LEAVING HIS house, I passed the dairy's fallen-down fences and close-cropped fields. In spite of the hard use, I could see that the soil in that part of the valley was fertile and dark. Last year's grass lay thick where cattle had not been let to graze it. The land seemed tailor-made for raising animals.

I liked Greg from the start because he is a hardworking, friendly man — the sort of person who springs to mind at the words "salt of the earth" — and because he seemed more willing than most ranchers to share space with difficult creatures.

But his place's beleaguered look gave me pause. Agriculture is a daily fight against entropy. Brace rails rot through and collapse. Water lines freeze and break. Fencing staples lose their hold on wood and must be hammered in again. Roofs turn swaybacked, then fall. Because a rancher maintains this balance between fixing and falling apart, something of his or her health, strength, and outlook is recorded in the condition of fields and fences.

Greg's dairy looked tired. Something about the sagging barbwire or gates tied together with baling twine reminded me of a car out of gas coasting down the highway. Momentum remained, but the driving force was gone.

Such depletion — which is epidemic across Montana and the American West — has several causes. The most immediate is economic: Agriculture, as practiced on small- and medium-scale family farms, does not pay a living wage. Poverty, or the threat of it, compels people to ask the soil for more than it can give.

Added to this are the effects of a hundred years of mechanization and increasing scale. Advised from all sides to get big or get out, farmers and ranchers have worked like mad to grow and modernize through the twentieth century.

In terms of efficiency, they were successful. I know a family near Deer Lodge, Montana, that bales more than six hundred tons of hay a year to feed their cattle. Such volume is not far above average, and many would say that it is still not enough to support a family. A cow can eat three or four tons of hay per winter in Montana's climate, and I was once told by a ranch consultant that a ratio of not more than one full-time cowboy to five hundred cows was necessary to turn a profit.

With numbers like those for benchmarks, it is no surprise that farmers like Schock cast their lot with big machines, synthetic fertilizers, herbicides, and genetically modified seeds. I do not blame Greg or anyone else for this. In the course of working on and managing ranches, I have been tempted to do the same.

There is a problem, though: A tractor is a demanding god, wanting offerings of grease and fuel. There are monthly payments to be made on new rigs and hefty repair bills on old ones. Implements — balers, rakes, mowers, cultivators, plows — have to be maintained and replaced when they fail. Each of these tools costs many thousands of dollars.

Even when things go well with this approach to agriculture, the eventual result is exhaustion. Greg's corn was a prime example. The crop was hard on him and hell on the soil. It required fertilizer, multiple applications of chemicals, and much tending. As the growing season was just barely long enough to bring ears to maturity, he had to worry over frosts, planting dates, and fickle weather. After talking with him and touring his place, it was clear that all his effort and expense barely kept the dairy afloat.

Driving out, I wondered if it was right or wise to build my grizzly fence on such a farm. Though the project, if successful, would help Greg keep more of his corn, I struggled to believe that it could improve his or the dairy's overall condition.

I came to the cornfield. Looking across acres of mud and last year's stubble, I knew that the crop should never have been grown so close to the mountains.

Staying out of it, I told myself, would not help the farmer or the bears. I turned my thoughts to what might be accomplished over the course of the summer. Smoothing away doubts, I began to plan the work.

THROUGH SPRING, Millie ate meat whenever she could get it. Waking from hibernation, she sought carrion with all her craft and strength. She needed calories and fat. Rot did not dissuade her. A putrefied carcass drove her to eat more quickly, which was less a matter of taste than of fear.

Many bears wandered in the Missions, each guided by habit and an excellent nose. Some, like Millie, were sows with newborn cubs. These family groups kept mostly to hidden drainages and remote snowfields. There were also young bears, three and four years old, recently cut loose by their mothers and learning how to make a living on their own. These adolescents wandered widely and sometimes aimlessly toward the far corners of the mountains. Some were Millie's offspring, and even if she knew them by scent, she never let them near her cubs of the year.

A sow with young is terrifying to encounter. Her righteous maternal rage is as volatile as gasoline. An ill-timed glance, a step in the direction of hidden cubs, is spark enough. Running afoul of a sow with cubs is a hiker's nightmare; to my way of thinking, that family group is the most fearsome thing in the mountains. If I knew of a sow grizzly moving through the hills with her offspring, I would walk miles to steer clear of her.

Other creatures see the matter differently. A grizzly cub, after all, is made of meat. In its first summer, a cub is not yet tough or fast enough to flee. Wolves and mountain lions know this, though they must be careful. Boar grizzlies know it, too. Being large enough to rout a sow, they among all creatures can hunt cubs with relative impunity.

But a determined sow is no pushover even for the largest, hungriest male bear. It's a rare boar that will risk his life for a meal—grizzlies do not reach maturity by making foolish choices—and it's a rare sow that will not fight to the death for her cubs. Still, Millie moved through the mountains with utmost caution. She turned away from other bears when she saw or smelled them, growing vicious if they came too close.

She hid in the vastness of steep country and was built for the work of surviving there. Flat grinding teeth backed her canines, her guts were iron, and a high hump of muscle made her paws into formidable excavators. Through April, she lived from carcass to carcass. Between such bonanzas she cropped elk sedge and fescue and flipped the glacial stones that littered high parks. Finding ant colonies, she consumed entire worlds. She ate the workers and the queen, the egg galleries and the dirt around them.

For a time, it was enough. Rich milk poured out of her, and the cubs grew stronger and faster. Having learned that their mother was salvation in a perilous world, they kept close. Even as they gamboled in the morning sun, they watched her movements.

Millie raised her nose to a southwest breeze and looked down to the valley's green grid of farms and pastureland. The season was advancing, and few winter-killed carcasses remained among the peaks. Things change quickly in the mountains, with feast becoming famine as quickly as clear skies give way to a storm. Though she ranged more widely each week, Millie mostly found bone piles or the leavings of other bears. Her stomach twisted and complained. Until it was full, she could not rest.

That is a bear's curse: Their waking season is short, and hunger rides them hard. In blooming spring, they think of winter. The world's plentitude must be consumed, stored beneath skin, larded away in preparation for the cold. The pressure is doubled for a sow that has spent energy on a pregnancy and must nurse her cubs.

In the valley below, blossoms covered the fruit trees. Alfalfa grew dark and tangled in the fields. Warm, light wind crossed the farms and ranches, carrying the scent of growth upcountry.

Millie's memory was very good where food was concerned. The smell of new growth, to her, was the smell of gophers denned in moist soil. It was the small heat of life fleeing through dark burrows, the earth peeling away beneath her claws, the easy crunch of rodent bones between her teeth.

All that was on the warm wind — and the smell of mankind, too. Noise rose faintly from the valley, an engine's buzz and the ring of metal on metal. Millie stood in the sun. She took a step forward and the cubs stopped playing; another and they stirred to follow. Slowly, with her usual caution, she began to descend.

3

Field

and

Fence

On the day I began building the electric fence at Schock's place, clouds hid the sun and the cornfield lay muddy and barren with stalks listing up in places from the ground. Midway along the eastern edge, a wire gate opened onto Hillside Road. I parked my laden truck beside it, opened the door, and stepped down to the gravel. My old heeler dog, rickety from years of chasing cattle, leaped down, stretched himself, and pissed in the grass. Glancing once at me, Tick began a thorough inspection of the wayside litter.

Westward across the valley, smoke was rising from the chimneys of farmhouses, pooling sluggishly in the air. I stretched in the road, letting the chill of a May morning brush dregs of sleep from my mind. Turning east, away from the field, I walked across the gravel and down through the ditch. A long day lay before me, but I wanted to see the trail before I started working.

"You've got to take a look," Greg had said when he gave me permission to start building the fence. "It's like a highway for the bears when they come down out of the woods."

I didn't take much convincing; ranch work has piqued my curiosity about trails. This is partially a practical consideration: When it comes to herding cattle through rough country, an understanding of faint paths means the difference between safety and danger, success and failure, efficiency and dithering. But my interest is also tied up with a passion for animal psychology. It springs from a conviction that I can better know the creatures around me by examining the marks they leave behind.

Coming to the roadside fence, I looked past its rusted, sagging wires and saw the trail: a green line of sprouted cheatgrass in a field of last year's tawny growth.

Perhaps a foot and a half wide, it bent in slow, even curves through old wheatgrass and brome. I followed the line toward the mountains until I lost it, then back until I was staring at my boots.

Every trail recalls its maker. Any animal that walks — a whitetail deer ducking into a daybed thicket, an elk straining up through loose dirt for the high country, or a person cutting the corner between car and office — writes something about its inner life on the ground. Depending on the species, the individual, and the day, that scribbling can speak of fear, grace, indolence, or a hundred other qualities. A cowpath's perfect adherence to the easy road — whether it is beaten deeply into floodplain mud or climbing straight toward a saddle — bespeaks the content torpor of the bovine mind. Braided elk tracks on a sidehill, narrowing and twisting across scree, argue the heavy, wary grace of the herd.

Until that day, the bear tracks that I had run across had been solitary imprints in snow, on logging roads, or in the wet squalor around cattle tanks. I had never walked a grizzly trail and was caught off guard by its look and feel.

The message was complex and foreboding. The trail was broad, its curves smooth and regular as they bent to accommodate hills and swales. It was a well-traveled, unhurried route — an artifact of animals that knew precisely where they were going, were moving at an unhurried pace, and felt no hint of fear.

There was something else: Studying the swerving line, I was struck by its familiar quality. It was more like a human trail than anything I had seen. It was a footpath, not a hoofpath, and elsewhere it might have led to a garden, an outhouse, or a place where people like to swim. *A foot like my foot,* I thought, and in my mind's eye, I could see the wide black pad of a grizzly paw pressing into the ground.

Looking toward the mountains, averaging the curves of the path, and taking a bearing straight down their middle, I found myself staring directly at a stand of aspens and dark pines — the timber bumping out from the mountain like the knuckles of a fist. Though it was a small unremarkable copse no more than half a mile wide, I knew Millie's Woods at a glance. As I looked across the distance, it seemed impossible that so many bears would favor such a small piece of forest. A few houses were visible — Greg's nearest to me and others hard against the edge of the woods — and I wondered how often the people who lived in them saw grizzlies.

Turning, I went to my truck and unloaded it: posts into one pile, crossrails into another, wire to the side. Taking a string trimmer, posthole digger, and six-foot rock bar from the bed, I looked to my work.

THE CORNFIELD already had a perimeter fence: a dilapidated five-wire affair that the bears climbed through as if it weren't even there. My plan was to ring this construction with a new one, stretching a trio of electrified wires around the rusted barbed strands. By building this way, I hoped that the old waist-high barbwire might dissuade a grizzly from thoughts of jumping.

Posthole digger in hand, I imagined how the fence would run and how a bear — perhaps an old boar that had been eating Schock's corn for years — might approach. In my mind's eye, he came after dusk when shadows could hide him and the moon was on the rise — padding down from Millie's Woods; passing through the big irrigation canal, which soaked him to the waist; waiting while the headlights of a solitary car cut the night; crossing the road with water dripping from his fur and darkening the gravel; coming finally to the bright line of wire and studying it carefully.

I could see him clearly as he peered at the uppermost strand, which gleamed in the moonlight at the height of his nose. Leaning forward, black nostrils opening and flaring pink, he smelled the tang of galvanization. He closed the distance to half an inch and touched the metal still lukewarm from the day. A sharp blue-white arc split the darkness.

Having been shocked by electric fences, I did not have to imagine the sound — which is like a brittle twig snapping — or the feeling. I know that for a creature in good contact with the ground — a damp-footed grizzly, for instance — anything over five thousand volts would hit like a hammer. The strike would pass through the bear in an instant. Every muscle on him would seize, a high and thin ringing would fill his ears, and the nerves of his face would burn. Shaking off pain, the boar would run for the hills.

The beauty of an electric fence is that, while extremely painful, its shock does no lasting harm. A modern energizer — the device that sends electricity along a fence's wires — is built to pulse, emitting jolts of current every second or so. Though the voltage gets quite high, the amperage stays low, and amps are what stop the heart.

The image of a bear testing the fence stayed with me as I choked and pull-started the trimmer. It remained while I knocked down a wide circle of last year's grass around the place where the first brace would stand. I liked the notion of a grizzly being driven away from a crop that might have been his undoing. *Back to the hills,* I thought, while the smell of cut stems and two-stroke exhaust filled my lungs. *Where he'll be safe.*

I MEASURED and marked the location of the first post, choosing a point two feet outside the line of the existing fence. Taking up the digger — a long-handled implement that looks like the overgrown bastard of a garden trowel and a set of fireplace tongs — I centered it over the place where I meant to make a hole.

I brought the tool up until my hands were level with the top of my head and brought it down hard. Feeling the soil give way, I opened my arms wide, spreading the handles so that the heavy curved blades drove inward.

The turf gave way with a sound like fabric tearing, and after stirring the digger through a wide circle to part the roots, I lifted the tool with sod pinched in its steel jaws. Light-brown clayey earth showed in the gash that I'd made. It was moist, fine-grained, and without many stones.

The digger could only take me so far. Every four inches or so, it struck a hard-packed layer or rang against cobble. Then I used the rock bar: chopping downward with its heavy blunt-edged blade, levering back and forth, shuffling around the hole, stabbing until the dirt was friable.

I dug until the pit's bottom disappeared in shadow, fetched a seven-foot-long post, set it into the hole, and used the rock bar to compact dirt around its bottom end. I took special care to seat the base firmly, standing back at intervals to make sure that the post stayed perfectly straight. Filling and tamping, I worked until the hole was full and the brace post set tightly.

It was good soil, and I moved joyfully along the line. When the sun fell down to the western mountains, I drove an hour home to Missoula, slept insensibly, and returned in the morning. Two weeks passed in clouds of dust and acrid two-stroke exhaust. A pain grew between my shoulder blades and ran out through the muscles of my arms. It was with me while I worked, and it followed me home at night.

"How is it going?" Gillian asked me at dinner after a run of long days.

"Feels good to build something with my hands," I said, showing her how my palms had blistered and begun to heal.

It would have been as true to say that I exalted in the labor, that the toil was simple and pure, and that it fed a part of me that had wasted in the absence of ranching. I was spending my days beside the mountains under Montana's flawless, endless sky, laboring at something that could save the lives of animals. It was a good job, and in late May, I worried that it would finish too soon.

BY MID-JUNE, it was eighty degrees before lunchtime, and the field stayed humid under a sprinkler's constant circling. That sprinkler, which is properly called a center pivot, is a modern, efficient, and expensive device for spreading water over crops with a minimum of daily human effort. Its long, straight arm of galvanized pipe extends from a fixed point at a field's middle and is supported at intervals by a series of wheeled towers. The towers, rolling ahead or backward on tall, heavy-treaded tires, trace concentric circles around the center point. Seen from the air, a pivot sprinkler can be identified by verdant growth and wheel ruts that look like ripples expanding on still water.

A pivot's spans rise high enough to accommodate tall crops like corn and sunflowers. Nozzles hang from them on thin necks of flexible tubing, dispensing water in overlapping fans.

Once installed, such sprinklers can be started, stopped, and moved with the touch of a button. The price for all that productivity and convenience is lofty. Schock's pivot, which was a seven-tower design meant to water ninety acres of corn and an adjacent patch of alfalfa, must have cost more than eighty thousand dollars.

The center pivot, being the tallest thing around and visible from every corner, dominated Schock's field. Its sound — the *chut-chut* of the end gun, the exhaling hiss of nozzles, and the groaning of drive motors as the machine crept along its circular path — was omnipresent in summer.

On a warm June morning, with the sun low in the east, the sprinkler looked as if it rode on a long white cloud of spray. I watched it as I eased my truck along the cornfield's southern edge, heading for a corner where I meant to build a brace. Gillian walked ahead, scouting for rocks or spans of old discarded pipe in the grass. With dark curly hair escaping her straw hat and cutoff shorts leaving a long span of browned skin above her boot tops, she was beautiful. She swung through the new grass, and I followed in the truck, passing posts that I had set in past weeks and braces that we had built together.

For the most part, I like to build fences alone. This has little to do with machismo or masochism, though I'm proud of what I can accomplish without help. I work by myself because I think most clearly without interruptions, and I dislike explaining things for which there is no fitting language.

But when it comes to building braces, I want company, because brace building is a dance. Not just any partner will do. A good one understands the process, the risks involved in each step of construction, and the unspoken needs of his or her counterpart.

Nobody starts out like that, and my first efforts with Gillian were slow and awkward. Though I had knocked together my share of H-braces — the bread-and-butter corners of ranch pastures — the cornfield required something called a floating foot. The design was new to me and relatively rare in our part of the world. Though Gillian was working as a teaching assistant on the University of Montana's demonstration farm in Missoula, she came to the task and tools of electric fencing as a novice.

Every five minutes, I had to pull a schematic from my pocket, stand back, and stare at the half-finished thing. Gillian watched, trying not to look skeptical. From time to time, she asked if there was something she could do. When we did spring into action, with the chain saw buzzing and hammers striking at all angles, I worried for our safety.

But after a week, things had changed. We built well and quickly, each anticipating the other's needs. When I knelt to stretch and splice, Gillian put tools in my hands before I knew that I wanted them. She knelt beside me, holding the wires in place, coming close enough that I could feel the heat of her skin.

As we worked, my eyes strayed to the mountains. The hills beside the Mission Range are unlike any others I have known. Where other peaks give way to sagebrush ridges or palmate basins, the ones above Schock's dairy retain an unbroken thatch of close-set timber. Most foothills are transitional country, belonging half to the domestic world and half to wilderness, but the ones beside the Missions have cast their lot entirely with the mountains.

Perhaps it is because of their beginnings, because the hills retain the whetted angles of glaciated stone. Beautiful but uninviting, they rise as obstinate as a palisade. My eye never wants to linger on them, has a tendency to skip past the thick dark timber, moving directly from the level valley to the gray peaks.

The gaze jumps over that bulwark of cedar groves and doghair pine, but the body cannot. To reach the top, a person must labor up one of a few rough roads or trails. Even strong hikers tire before the mountains open for them, and the hills see little human traffic. Their ruggedness is a godsend for bears, as it lets them pass secretly along the valley's margin. Having seen telemetry data from the bears that Stacy caught and collared, I have come to appreciate how easily grizzlies cross steep country. Often, it looks as though they travel over mountains faster than they do flat ground.

In 2006, for example, during the summer after Stacy released her, Millie went on a walkabout. Leaving the valley floor from a spot near Schock's field, she climbed nearly to the ten-thousand-foot summit of McDonald Peak. Cutting northwest over sheer ridges, she descended into the Swan Valley and made a fifty-mile loop near the town of Condon. Looking at her path, I am struck by both the distance that she covered and her apparent disregard for topography. She visited mountaintops as if climbing were no trouble, stopping in for whitebark nuts or army cutworm moths as casually as a commuter might swing by a corner store. Knowing where she walked, I cannot see the pinnacles and deep valleys of the Missions without marveling at her strength.

AROUND THE end of June, while Gillian and I built braces, Millie left the enfolding foothills. The sow descended with duff shifting light and dry under her paws, her cubs loosing small landslides behind her. The valley below was flecked with houses, and the sounds and smells of humankind grew strong as she passed above the ends of east-west roads.

North and east of Red Horn Road, she followed a peninsula of timber, keeping to cover and stopping often, until she came to the wide earthen irrigation canal running along the western margin of the mountains. Crossing this humble Rubicon, drinking midstream, she entered her namesake woods.

From far above, that place is a rough-margined dark teardrop — thin end reaching east to forest, front belling out — with farmland on three sides. Scattered houses flank the north and south fringes of the timber, but the heart of the woods is whole. Except for birdcalls and wing beats, it is quiet. Save for the occasional movement of spooked deer, it is still. Through a happy accident of morphology, water is never far from the surface. Seeps and pools abound, with aspens and tall ponderosas overspreading all. When summer burns the rest of the Mission, Millie's Woods remains cool. Sunlight, scattered by leaves, flecks the ground like confetti after a parade.

She came into the woods seeking shade, water, and undergrowth in which to hide. She stayed to raid north into hayfields and pastures, searching for clover and ground squirrels. Along with many other animals — coyotes, deer, moose, and elk — Millie favored the woods because it was a last outpost of wilderness and a beachhead in the domesticated world.

She knew, too, that whitetail fawns lay in the thickets. Cautious but driven by hunger, the does played an ancient game of hide-and-seek among the alder stands and aspen trees. Stashing their offspring in unlikely places, they browsed on tender growth.

Fawns are born knowing the game and are particularly good at keeping silent and still. Camouflaged by coats that look like dappled bark, they stay motionless until a person all but steps on them.

Walking, I have blundered upon and stood over such newborns. They are always settled nose to tail, curled as tightly as they were in the womb. They never blink, bolt, or move except for their panting ribs.

On the early side of dawn, Millie brought her cubs along one of the many narrow trails that filigree the woods. Coming to a small clearing, she smelled the milk-sweetness of a hidden fawn. Her broad black nose pulled steadily at the air, drawing her on.

The bedded fawn heard stems and leaves soughing and the sound of breath drawn and exhaled. Its small world was a ring of sheltering grass and a high vault of leaves. Millie's head and shoulders entered that circle, as wide and dark as the trunk of a pine.

Without hesitation, she set a heavy paw on its neck and shoulders, pressing it into the ground. Small hooves threshed air, and the fawn bleated bloody murder. Millie leaned in, and the deer leaped against the first bite's pain, springing up with desperate strength. An unceremonious blow brought it down.

The arriving doe trotted in wide circles, dodging as close as fear allowed. She made coughing noises that carried through the trees, agitating other deer within earshot. Millie took no notice, and as morning brightened above McDonald Peak, the doe left the grizzlies to their meal.

BY EARLY July, Gillian and I had finished the braces along the east, south, and west sides of the field. Constructing a homemade spooler from plywood and scrap pipe, I turned my attention to the solitary work of stretching wire.

On flat, open stretches, I mounted the device in the truck bed, speared three half-mile-long coils on its spindle, fixed the wires to a brace, and eased ahead. Looking in the rearview mirror, I watched the spools spinning fitfully as they played out galvanized threads. Driving along the fence that way — letting the pickup do the work — filled me with an immense sense of progress and satisfaction. Walking briskly alongside, Tick seemed to share the feeling.

Arriving at the field's southwest corner, I could tell that the going would be harder. The field's west side bordered a neighbor's place, which meant that I couldn't drive outside the barbwire. Within the field, irrigation had soaked the bare soil into an unnavigable morass.

From then on, I pulled the wire by hand, straining like a horse in a harness and slipping in mud. My progress slowed, and the corn seemed to grow taller each time I looked at it. Before I had completed the east, south, and west sides of the field, the crop stood nearly to my navel. The stalks were swollen and rigid with water, and when I cut a corner on my way to or from the truck, I had to take care not to break them.

Bears were increasingly on my mind. The crop had grown tall enough to hide a grizzly on all fours, and though there were as yet no ears on the stalks, summer was passing at a disquieting pace.

"Second half of August," Stacy had said when I asked him when bears started coming to feed on the corn. As I watched the crop grow, it seemed impossible that the grizzlies would wait so long. Nervous about whether I could finish the fence in time, I saw Greg driving the road one afternoon and flagged him down to talk.

He leaned from the window of his around-the-ranch rig, a nineties-vintage teal Geo Tracker with an outlandishly large hood ornament in the shape of a wide-winged swan and windows that hadn't closed in years. When Greg wasn't using the car, it sat in front of his driveway with the driver's door open rain or shine. I used to pass his house and wonder why he left the door that way until I came to understand that the rig was the sovereign property of his chocolate Lab, who used it as her kennel and den.

"Anna," he said to the dog, as she stood barking and wagging in the shredded back seat. "Shut up."

He turned to me. "What were you asking?"

"If you've seen bears around," I said. "And when the corn will get ripe enough that they'll start coming to the field."

"Anna!" he shouted over his shoulder. "Shut up!"

Evidently pleased, the dog yapped vigorously.

"They're always around. Last night a sow and cubs came out of the woods and headed north toward Post Creek. As far as the field goes, it depends on the year. Sometimes they don't come until the end of August. But this summer, well, you see how far along the corn is."

I followed his gaze. Corn stood chest high, its first tassels swaying in a light wind. Leaves stretched up and out, bending into delicate arcs under their own weight. No ears had started, but the corn had a new strength and uniformity. With every stalk shouldering its neighbor and growing higher by the day, the field had a sense of pregnant readiness that I had not noticed before.

"A good year," Greg said. "Look at the apple trees. Look at the water in Post Creek—*Anna!* Quiet! *ANNA!* Rain when we need it, hot weather like we've had. Year like this—we might see the corn ready and bears coming a little sooner. Maybe the first of—first week of—*ANNA!* August."

That gave me two weeks to finish, and I stood considering the amount of work yet to be completed. Taking note of my silence, Greg asked if I thought I'd have the fence done in time.

I told him that I would, but saying it left me uneasy. Turning back from the field, I saw that Greg's expression had grown serious.

"Careful out there," he said, and Anna quit wagging her tail.

"Few years ago, I had this little rig"—he patted the cracked dashboard with a heavy hand—"out in the corn. It was late August, and I'd gone in to fix something—a broken nozzle, maybe. Bears were in there, plenty of them, and my plan was to drive out through the field and work beside the car so I could jump back in if I had to. Down I come along the access road toward the center of the pivot, but just before I get where I'm going, the Tracker bogs down in a wheel rut. You know how it gets when there's water in that ground."

I nodded, remembering the way the thick, heavy mud had weighed down my boots in spring. The dog barked once from the back seat.

"Anna," Greg said softly. "Quiet."

"I was stuck tight. Four-wheel drive didn't break me loose, and there I was, dead in the middle of that field, with the corn head high and ripe. Couple nights before, I counted nine bears going in at dusk.

"I took four or five tries at breaking out but just spun my wheels and crawled sideways along the pivot rut. Pretty soon I'd slid off the access road—which wasn't much wider than a car—and had the hood buried in corn."

"So, what'd you do?" I asked. "Call somebody to give you a pull?"

"Nobody around to call."

"Walked?"

"Walked? Hell, no. I don't walk in there, and you shouldn't, either. Not when it's up and the bears are around. No, I don't walk in there, not for nothing. I put the car in low and crawled my way down the pivot track with all four wheels spinning and mud flying, mowing down stalks. Drove like that to where the corn ended and the ground firmed up."

Looking past me, Greg seemed to be mulling something over. I waited for him to go on, supposing that he must have knocked down more corn that day than a hungry bear could topple in a week.

"Made a mess," he said, apparently thinking along similar lines.

Something new was in his eyes, a mix of fear, wonder, and respect that I had not seen before. It was the look of a man who had lived long among grizzlies, knew their strength, and had learned an animal caution from years of proximity.

Greg watched me closely.

"Tell you this: I'd knock down a pile of corn before I went walking in that field in August. Like I said, you don't go in when the corn's up—there's no reason good enough."

He rattled the Geo to life, provoking a high-pitched salvo from Anna.

"Better get on," he said. Splitting his craggy features with a smile, he headed down the road in a cloud of dust and frenzied barking.

HE LEFT me to the task of clearing a route for the field's north-side fence, which was to follow the course of a steep-sided ditch. Water flowed slowly in that channel. Slough grass, cattails, and teasel grew shoulder-high.

I stumbled along the bank with the cornfield on my left side, a dense stand of alfalfa to my right, and Greg's old barbwire fence almost underfoot. On the north side, that fence was in particularly poor repair. Dark creosote-soaked posts leaned at varying angles, with rusted wire sagging between them in arcs.

In a farmer's perfect world, a pivot sprinkler makes a full circle, watering the crop as a clock's hand moves through the hours. Schock's field, however, takes the shape of an irregular square with a missing northwest corner. To cover it, the sprinkler has to work clockwise around its fixed center, move through 270 degrees of the compass, reverse its direction, then roll back the way it came. Things are further complicated by the fact that one pivot waters two crops — corn and alfalfa — and must cross the fence that exists between them.

Walking that fence, which was the cornfield's northern boundary, I saw that the barbwire had been cut and folded back in places to accommodate the pivot's movement. There were six crossings — one for each of the large-wheeled sprinkler towers — and in each gap, I could see deep tread marks left by the tires as they crossed from the cornfield into the alfalfa and back again.

The crossings presented a difficult problem: I couldn't enclose the alfalfa field in electric fence, as that would add greatly to the cost and scale of the project. Neither could I leave openings, as Greg had in his old barbwire fence, for grizzlies to walk through. The pivot sprinkler — nozzles, towers, four-foot-tall wheels, and all — would have to pass over or through my fence without diminishing the electric current in the wires.

I had two choices: I could build a series of twelve braces — one on either side of each tower crossing — and invent a hinged, spring-loaded metal gate to hang between them, or I could come up with some way to let the wheels drive over the wires. Measuring the corn's growth against the work yet to be done, I chose the latter, more straightforward course.

THE LAST push of fencing was a race and a battle against tangled, seething fertility. Knots of grass sent me sprawling. Pliers disappeared in the weeds, and teasel thorns split the weave of my jeans. A truck could not go far in there, so I had to pack tools and materials from one place to another on foot.

One day toward the end of July, I looked closely at the corn and, with a feeling just short of panic, noticed swellings at the bases of the leaves. A week later, I crossed the fence and picked a sprouting ear. Inside was only fiber, but I knew the bears would give no quarter when the kernels grew. The thought of them waltzing through an unfinished fence spurred me on.

The tower crossings were my last substantial challenge. Knowing that steel wire has a surprising capacity to stretch, I decided to leave some slack in the strands, use flexible posts at either side of each crossing, set large rubber mats beneath the wires to keep them from coming into contact with the ground, and let the pivot drive right over the fence.

I tested this strategy on the outermost crossing, and it worked like a charm. The sprinkler crept ahead, its electric drive motors whining until tread met wire. Then the fence flexed downward, growing taut as it descended. It took two minutes for the tires to pass, and all that time I waited for something to snap.

Nothing broke. The machine moved on with its deluge, and my fence sprang back into place. Going to it, I found the wires undamaged and marked with a thin tan coat of mud.

It was the last crossing that tried me. There, closer to the sprinkler's center point, the wheel track bent into a tighter circle and the fence passed over an earth-covered culvert. The lay of the land thwarted me, dipping and swerving in a way that made it impossible to set the fence at right angles to the sprinkler's path. Testing it out, I found that the wheels could do a lot of damage when they crossed the fence obliquely.

With time growing short, I settled for lowering my top wire by a few inches to be sure that it wouldn't be torn loose. Setting the rubber mats in place below the fence, my mind was far from easy. The wires seemed too close to the ground to stop a bear, and I left the place under a double load of tools and worries, feeling that my work there was unfinished.

JULY WAS fully exhausted before I managed to rig and crimp the electrical connections, sink the long galvanized rods of the grounding system, and install a brand-new energizer. After finishing the final splices, I switched on the current and looked across the wire. The corn had risen beyond head height, becoming a wall of leaves and stems. Though the ears were not yet ripe, I could see them growing and could well imagine how they would call to the bears. Against the draw of such a feast, my fence looked unsubstantial. The highest of its three wires did not quite reach the knee of my blue jeans, and when I nudged a composite post with my boot, the whole construction shivered with motion.

Greg was moving cattle down Hillside Road with his son, and I watched them come abreast of the cornfield. Holsteins crowded against one another, bunching while the men revved the engines of their four-wheelers and pushed the herd from behind.

The drovers were not unkind, but they were in a hurry. When the lead cows balked at an open gate, Greg and his son shouted and pressed. A knot of cattle was soon edging toward my new fence.

The outermost cow was shoved and shouldered backward. Ten feet separated her from the fence, then five, then her hindquarters were driven against the wire. Nothing happened for a moment except for a post flexing under her weight. No more than a second could have elapsed, but I had time enough to wonder if there was a short circuit somewhere or if I had chosen an underpowered energizer.

I was on the verge of dismay when it happened: With a high yowling bawl, the cow leaped forward. She cut a swath across the road away from the fence, scattering calves as if they were bowling pins. Other cows followed, and soon the herd passed through the open gate with Greg and his son behind.

I leaned down and touched a voltmeter's metal probe to the uppermost wire. The display pulsed at one-second intervals. It read 8.5 kilovolts — a hell of a strong charge.

4

High
Summer

I N T H E Mission Valley, August raced ahead without me. Choke-cherries ripened, blackening on limbs beside ditches, streams, and reservoirs, flourishing in the shelterbelts of abandoned houses. Birds chattered in the leaves. Bears visited the fruit trees at night and ate standing up, raking with claws, pulling branches through their mouths, swallowing pits and all, resting bloated until they could resume the feast.

Plums swelled and softened. The grizzlies knocked hot-fleshed fruit to the ground. They sucked hard at the teats of summer, eating until their guts succumbed. Apples ripened everywhere — Bitterroot McIntoshes on leggy, untended trees; prized Honeycrisps in orchards; crabapples on the banks of creeks and sloughs — and the feast became frenzy. The grizzlies fed in a series of massive inhalations, taking food the way a skin diver breathes before ducking beneath the waves.

This was the rule of August: During the day, the valley belonged to humans. Tractors worked the fields, and children played carefree in the yards. People swam in shady eddies and picnicked beside the creeks. Life, even at the edge of the woods, went on as if ravenous creatures were not hiding among the trees.

At night, the bears came out. With twilight's coming, bears stirred in well-hidden daybeds. Stretching in the cooling evening, they made sorties from Millie's Woods and other sheltered places. The valley belonged to them after dusk, and people shut their doors. Yard lights spread yellow circles on the ground, and farm dogs patrolled the edges, barking into the dark.

The valley floor may be perilously full of humans, but it's lavishly stocked with food. Grizzlies eat a lot of domesticated fruit in the Mission, more than in any other part of Montana. They've been making use of orchards for a century, and their scats have spread seeds across the landscape, raising countless feral trees.

The bears that came down in August knew every corner and fold of the land, each tree and unfenced garden. They were careful, if not afraid. Though they went nightly among houses, they were very seldom seen.

Millie taught her daughters how to forage in this risky cornucopia. The little bears were compact, pretty creatures. Only a few months old and full of restless energy, the cubs aped their mother's movements: When she dug in the dirt, they dug, too. When she paused midstride to scent the wind or listen, they stood still.

Coming down from Millie's Woods, they passed near enough to Schock's dead pit to sniff at bones. Crossing Hillside Road, Millie and her cubs saw headlights coming, heard rising engine noise, and loped away through the alfalfa growing north of the cornfield.

Avoiding houses, they made nocturnal rounds of places that she knew well. Visiting familiar trees, Millie kept far from lights. Though the cubs had grown, she guarded them jealously against the world. Roaming the valley for more than a decade had steeped her in caution. Millie became, as all grizzlies do at night, a heavy smooth-walking shadow. She was a faint noise from without, a suspicion in the minds of men and women peering from doorways.

She was efficient, hungry, and surprisingly discreet for an animal of almost five hundred pounds. The cubs understood enough to eat the way their mother did. Gorging themselves sick, they rested and began again. A swaddling of fat grew under their skin, and their hair showed glossy in the moon's white light.

I PASSED the first three weeks of August without visiting the fence; another project took me out of state and deep into the backcountry of Idaho. Steve had volunteered to drive up from his stomping grounds in the Madison, Ruby, and Big Hole valleys to check on things in my absence. Gillian planned to visit the field to read the voltage in the wires and look for evidence of bears. Though things were left in capable hands, the cornfield was never far from my mind as I hiked in the White Clouds Wilderness.

When I finally came down from the mountains, I switched my cell phone on and watched messages flood in. The first email that I received from Steve made it clear that there had been trouble at Schock's field.

The issues, he wrote, had started at the spot where the pivot's sixth tower rolled over the fence. There, where the lay of the land and the angle of the fence had forced me to drop the uppermost wire by a few inches, a bear had jumped across. Steve had found tracks on the rubber mat beneath the wires — big hinds and fronts set tightly together, as though the bear had gathered itself for a leap. He thought that he had seen the same tracks leaving the field — though he couldn't be sure that both sets of marks had been made by the same bear.

Hoping to fix the problem, Steve had added a pair of electrified mats to the sixth crossing. These new additions — six-by-eight-foot squares of rubber horse-stall flooring topped with electrified mesh — were intended to keep grizzlies from walking anywhere near the wires.

A subsequent note from Gillian made it clear that the results had been far from perfect. She wrote that the wire voltage had mysteriously dropped and bears had gotten into the corn. Whether they remained in there or not, she couldn't say.

In late August, the sun stayed long in the sky, and the Missions were casting off their last snow. The sky was powder blue, with forest fires hazing the distance. Towering clouds bore up from the southwest, gravid with rain and lightning. Grasshoppers rose from the ditches along Red Horn Road.

Driving with the window down, I could hear their chirruping above the engine noise and rattle of tools in the bed. Flying up from the grass, the insects moved over the gravel like old leaves in a wind. They beat softly on the hood and windshield, bouncing away or bursting into chartreuse blots. Sometimes Tick, who rode in the jump seat with his head out the window, snapped at a hopper as it passed his nose.

I arrived at Schock's field with a storm of bugs in front of me and dust behind. As I slowed to make the turn from Red Horn onto Hillside Road, one of the hoppers soared in through the window, came to rest on my arm, and remained. Glancing down, I saw the creature's yellow tiger-striped carapace, slate eyes, and barbed wrong-way-round shins. Though small, the grasshopper was substantial. I could feel its weight and the sharp pricking of legs as it turned on my arm.

Coming to the start of the cornfield fence, I slowed the truck and steered for the road's margin. Giving my skin a final needling, the hopper leaped away, sailed out the window, and spun through the air. Like an engine guttering, choking, and catching, its wings spread, paused, and blurred to a streak.

Stopping the truck, I watched the hopper's flight. It followed the road, keeping above the grass and settling finally on a seed head. The grass was smooth brome, and its tall slender stem drooped under the creature's weight. Bending double, the stalk flexed toward the road. The hopper held fast, and as it came to rest, I saw that the whole scene — bug, seed head, and bowed grass — hung inches above a pile of bear shit.

I stepped from the truck with Tick at my heels, and even before I came to the scat, I could feel the change that had come over the field. Though I had been gone for less than a month, the place seemed foreign and strange. The corn had grown very tall. Most stalks reached seven feet or more, and the stand had thickened until each plant's leaves tangled with those of its neighbors. It looked like a thicket, not a planting, though toward the ground I could still see the precise linear arrangement of the stalks, each with its clasping hand of roots.

The stand had changed color from bright new growth to forest green. It gathered shadow near the ground, and darkness mingled with the leaves.

Walking to the pile, I studied it. Tick looked on, sniffing mightily at the air.

Plucking a length of grass, I bent to one knee. The scat looked fresh, not more than a few hours old. Studded with hard, round masses, it formed a mound eight inches across and four inches high.

Working with my stem, I freed some of the bigger bits and flicked them aside on the gravel. I found plum pits, apple seeds, and chokecherries in great quantity. Here and there I recognized the tangled remnants of grasses. *A fruit-eating bear*, I thought, taking comfort in the absence of corn kernels.

Standing aside from the blunt, too-human smell of omnivore scat, I looked north. Not far away, near the place where the old bear trail came out of Millie's Woods and crossed the road, I could see another pile. Walking over, I found that it, too, was composed mostly of stone fruit. Other dark mounds, five in all, dotted the road at intervals.

Looking west across my fence to the corn, I saw a sixth heap in the barren dirt beside the outermost row. Even from a distance, I could see that the pile's color was lighter than the others.

I called Tick to me, feeling suddenly alone and exposed. He came at a run and waited while I weighed fear against my desire to cross the fence. When I finally stepped over the uppermost wire, taking care to clear it by several inches, my dog stayed behind. He paced up and down along the line, having been shocked too often to venture a jump or a dive.

Ten paces separated the place where I entered the field from where the bear shit lay beside the corn, and I heard the low, hot sound of my heart as I walked. The crop loomed up until I could see nothing but the stand, which had the aspect of a wave prepared to break. Warnings flickered up from the deeper parts of my mind: *A bear, very near, and you cannot see.*

The thought became a fearful koan uttered inwardly with every step and gaining force with repetition. My urge to flee became a physical pressure. It was something like magnetism, as though a repulsive force emanated from the stand.

Soon I stood over the heap, looking down at a mass of darkened kernels—some masticated, many entire. The heap's uniformity was striking. Where the other scats revealed a mixed diet, this mound told the story of a creature that had been eating only corn in great quantity, bingeing day after day on the crop.

But it's old, I thought. I toed the stuff with my boot, and the dry, hardened mass moved as a whole. Standing over it, I dared to hope that the grizzly had quit the field.

After listening for a time, I searched the ground for information.

Nothing stirred but an easy southwest wind that set the leaves and tassels moving. Even a human nose could recognize the soft, full fragrance of kernels on that breeze — a smell of plentitude, fecundity, and increase.

The sweet air reached Tick where he stood outside the fence. He pulled eagerly at it, shook himself, and trotted a few steps down the line. *Very near, and you cannot see.* Repeating those words, I backed away from the outermost stalks.

At the truck, I resolved never again to walk in or near the field without bear spray clipped to my belt. I knew enough to respect that nine-inch-tall red-and-black canister full of capsaicin and propellant. Years ago, I was in the cab of a pickup when a can of the stuff sprung a leak. I took no more than a lungful of the red-tinged peppery vapor and wasn't directly sprayed, but still it felt as if a giant mouth had been placed over my own, that it was sucking my breath away and keeping me from drawing another. I remembered how, after I had crawled from the truck to retch and heave in the road, the pepper had continued burning my lips, throat, and eyes.

The spray came in a small nylon holster from which it could be slipped quickly at need. Directions were printed on the can's side in small text, but I did not have to read them. The steps were seared into my memory, and I rehearsed them as I put the truck in gear and headed north along the fence.

Out of the holster, I told myself, *at the first suspicion. Off your belt and into your left hand, and if you see the bear or hear it coming, slip the safety — that small tab of glow-in-the-dark plastic — with your thumb.* In a decade of living and working among bears, I had never needed to take the process further. Still, I reminded myself of the rest. *If the bear charges, wait until it closes to thirty feet, which will happen in the blink of an eye; press the can's plunger; and*

watch a red cloud fly out. Two seconds, aiming low. Another blast if
the bear doesn't turn. Seven seconds to an empty can.

BY THE time I came to the field's northeast corner, the sun was well
overhead and bright in my eyes. Where my electric fence turned
from the road and struck westward, I parked the truck and took
a voltmeter from the toolbox. Made of green plastic and brushed
metal, the meter was a new model that could read voltage and amperage and turn the fence off and on with the touch of a button.
This last function let me test and fix sections of the fence without
walking back to where the energizer hung at the field's corner. The
device was like a dowsing rod that pointed unerringly toward trouble. It saved me miles and hours of walking, and I prized it.

Hanging the voltmeter opposite the pepper spray on my belt,
I walked to the fence and stooped to measure its charge. I held
the meter's probe in place until blocky digits flashed across the
screen, reading two and a half kilovolts. Suspecting some mistake, I
pressed harder against the wire. The readout remained unchanged,
displaying the same dismally low number. I tried the middle wire
and the lowest one with even worse results.

When I had left for the mountains at the end of July, the fence
had been running three times as much current. No spot along
its length had showed less than eight kilovolts, and in places, the
meter had topped nine. Eight kilovolts were enough to deliver a
stopping shock, well above the six thousand volts necessary to turn
a curious bear.

Two and a half thousand volts, though, was more slap than
sucker punch. On par with a jolt of static electricity leaping from
a doorknob, the fence's strength left much to be desired. It would
not stand in the way of any determined animal with late-summer
hunger in its belly and the promise of ripe kernels in its nostrils.

There has to be a fault, I thought. Only a break in the line or metal touching the wires would draw the current down so far. I checked another value on the meter. Amperage, the direction and measure of current passing a given point, was displayed on a corner of the screen. The machine showed twenty amps on the move, with an arrow indicating that the current was flowing westward.

Knowing that the arrow pointed toward a place where electricity was leaking from the fence like water from a cracked pipe, I went to the truck and loaded a bucket with tools, setting the wire stretcher in, then crimps, pliers, and insulators. Unwinding thirty feet of high-tensile wire from a spool in the truck bed, I bound it into a tight roll and started down the line.

It was slow going. The grass stood waist-high and thick, and its stems bent and parted around me like a bow wave. As I walked, the pockets of my work pants filled with seeds. My mind stayed uneasy, and my eyes flicked back and forth between my work, my feet, and the near wall of tall corn. The stalks stood no more than thirty feet away, and I could not shake the vision of a bear emerging at full stride.

Could you get to the spray in time? I wondered. As quickly as I could, I dropped my spool and bucket, fumbled with the Velcro closure of the holster, freed the bear spray, and thumbed back the safety. All the while I counted out the seconds, *One, one thousand; two, one thousand; three, one thousand; four* until I stood with the canister pointed at the stalks.

I knew that grizzlies moved fast — that they could rival a quarter horse at a dead sprint, which put their speed at around thirty miles per hour. Gathering my things and walking on, I tried to work out how many feet per second that was. Getting the right answer seemed important enough that I pulled out my cell phone and punched in the numbers. *Forty-four,* I concluded, realizing that the hypothetical bear had beaten me handily.

I struggled through the grass until I came to one of the places where the pivot sprinkler crossed the fence's line. In circling, the tires had left sodden bare ruts in the fields, and these gave way to muddy tread patterns on the rubber mats that I had placed beneath the fence. There were other marks, too — a haze of dirty smudges that might have been old prints or bits of mud fallen from the passing sprinkler.

I examined the smudges for a while but could make nothing of them, as the water had obliterated all detail. Finding no issue with the fence, I carried on.

After a quarter mile, I came to the sixth crossing, the place that Steve had modified in my absence. It was clear that he had been working in a hurry. Steve, as a rule, is a meticulous man with a temperament running toward tinkering and perfection. Here, though, haste had leant a disorderly cast to his efforts.

Two electrified mats, "unwelcome mats," as he called them, now guarded the crossing. Both made use of rubber horse-stall pads to insulate their metal elements from the ground. Atop that rubber, one mat bore a large segment of hog wire — a welded grid of thin steel rod. The other was covered with a snaking nest of three-eighths-inch cable, which looped in circles and was fastened in place with zip ties. Short spans of wire reared up from the panels, arcing out like antennae, and Steve had fastened colored tape to the ends, warning people against touching them.

At first glance, I thought that the mats had been pitched haphazardly on the ground. Soon, though, Steve's intentions became clear. The cabled mat had been laid across the pivot track, and I guessed that he had built and placed it to withstand the punishment of the tires. The other mat lay off to the side, straddling a bear's most likely approach.

Both mats were cupped in places with the shape of the land, and they held water in still, mirrored pools. Behind me, the pivot worked on through the alfalfa field. Watching, I could see the sprinkler coming my way, rolling ahead by a foot or two per minute. Turning again to the mats, I noticed that their rubber surfaces were dotted with dark specks. At first, I took the marks for bits of gravel. Bending down, I could see that they were the corpses of pill bugs, grasshoppers, flies, and spiders.

Small bodies, washed gray by the sprinkler's passing, flecked the mat's entire surface. They lay thickest at its edges and in the places where heavy railroad spikes had been hammered in to hold the mats fast to the ground. Spent bugs were concentrated into piles and rings around the big nails, so each spike head looked like the bull's-eye of a target.

Supposing that I had found the source of the fence's short, I knelt and took voltage readings from the head of a spike, the hog wire, and the fence. I found that the mats leaked electricity, grounding enough current to kill an insect through the railroad spikes that held them in place. But they weren't spilling enough to account for the whole drop in the fence's charge. With twenty amps of current flowing farther down the line, it was clear that the main problem waited to my west.

I shut the fence off for long enough to pry the spikes from the ground. Straightening and stretching my back, I peered north to where the sprinkler watered alfalfa. Taking a step in that direction and stumbling on an unseen stone, I looked down and saw the track.

The pivot's curving rut held a perfect print. I saw the meaty forepaw — as wide as a salad plate and shaped liked the ball of a human hand. Ahead of it, the grizzly's toes were recorded as soft ovals, as though eggs, held lengthwise, had been pressed into the fine-grained mud. But the claws were most striking; they looked at odds with the rest of the track. In contrast to the barefoot softness of heel and toe, the claws left sharp black pricks. They seemed too distant from the pugmarks — separated by at least two inches. Had I found the track elsewhere or known less of grizzlies, I would have supposed that the thing was a prank — the mark of a human hand amended with the claws of a monster.

The sprinkler crept onward, piddling. Before dawn, it had circled to the alfalfa field's far side and reversed back toward the corn. Looking down at the track's immaculate preservation, I knew that the grizzly had come after the water. The mud held every fold and feature of the pad, and the claw slashes retained a crispness that could not have survived soaking.

There is a bear, very near. Studying the land for sign and finding none, I decided that the grizzly had crossed rather than followed the rutted track of the tower wheels. That a bear would walk parallel to the fence instead of toward it seemed to offer some hope, and it cemented my resolve to get the wires fully charged again.

Pushing on far from the truck, I came to a sward of fuller's teasel. Earlier in the summer, the stalks had been green, and gentle long-nosed bees had made the rounds of the purple florets. Through August, the plants had cured into thorny brown maces that snatched at my pants and boots. Both hands full of tools, I lumbered on.

I fought stems for twenty paces before the weeds, all at once, released their grip. My legs swung easier, and looking ahead, I could see a dark interruption in the growth. A linear, curving shadow ran parallel to my fence about ten feet outside of the wires. Following it and peering down at my feet, I could see the trampled wreckage of stems and parched leaves.

I made use of the easier going, swinging along until I emerged from the last of the teasel into grass. There I stopped short, with a recollection of Stacy's voice ringing in my ears. "Trails like highways," he had said when we first spoke about the Schock dairy.

Ahead, the dark line that I had followed in the teasel became a wide path through brome and timothy — a bear road in regular use. Peering along the track, I could see where grizzlies had departed from the well-traveled way. High grass recorded the mark of every paw. Shortly, I came to where an animal had tried its luck with the fence. A swath of flattened rye — all the stalks running together like wood grain — showed where the bear had splashed through the ditch, come to the wires, tested them, and turned back. The whole effort left a mark like a wide uneven C, and the path by which the bear had fled from seemed straight enough to indicate haste.

I walked another quarter mile, passing a dozen places where bears had crossed the ditch. Studying the telltale crimped stems, I searched for and found the paths by which bears had approached and fled from from the fence. As I went, I attended carefully to the wires, clearing them of trampled grass and weeds.

By the time I neared the field's northwest corner, I was satisfied with two things: first, that no grievous electrical fault existed to my east; and second, that the bears whose tracks marked the ditch grass had been turned back by the new fence.

At the corner, the main branch of the bear road left the cornfield behind, striking northwest toward the thickets of Post Creek. A smaller spur crossed the ditch, running straight toward my fence. To my surprise, it continued on the far side as though the three electrified strands didn't exist.

Beyond the fence was a fifty-yard-wide patch of alfalfa. Greg, being resourceful and efficient, grew feed in the unharvestable corners of his field. Peering across the low flowering alfalfa plants, I could see a line of downtrodden foliage. It pointed straight toward the corn.

But looking closely, I grasped that a bear had not broken into the field but had forced its way out. Beyond my own posts and wires, the old fence — built there of woven wire — had been hit with great force from within. Its strands had been split and the broken ends bent outward.

It put me in mind of the aftermath of a highway accident — the wreckage left behind when a car careens through a guardrail, bending it until the metal buckles and folds. The old fence was careworn and had been patched many times. It was a place where animals had been breaking through for decades and where Greg had labored at various forms of repair. Splices and patches formed steel macramé, and looking carefully, I could see one kinked strand reaching out beyond the rest. Following the line of that narrow finger, I saw where it met and grazed the lowest wire of the electric fence.

A light touch, but it was enough. Taking a knee, I set my voltmeter to the broken wire and confirmed what I all but knew: A large amount of current was passing from Greg's old fence into the ground.

Stretching and mending took longer than I thought it would. Every time I pulled on a wire, an old splice gave way. As I worked, I tried to puzzle through all the evidence at my disposal. The best theory I could come up with was that a bear had gotten in at the sixth crossing, like Steve had said; lived in the field long enough to shit a pile of corn by the main road; then come out this way, shorting the fence as it fled.

I could not be sure of this or certain that all the bears were gone from the field, but I hoped for it as I fixed Greg's old fence, bending bits and pieces into place until every wire end that was sound enough to stand a twist had been joined to a mate. I worked carefully, supposing that if a bear had gotten out without much of a shock it might try to come back by the same route.

Switching the fence back on, I found the charge restored to an acceptable six kilovolts. Glancing around, I saw that the afternoon was starting to wane.

The westward-tipping sun burned hot and yellow. The day was windless, stifling. Grasshoppers kept up their maddening noise all around. Looking at the sun, I grew suddenly and painfully aware of my thirst. Distracted by all the bear sign, I had forgotten to drink at the truck or bring my water bottle. My tongue lay thick and rough in my mouth, and a thought came unbidden into mind: *It was shady in the corn.*

There is a bear, very near, my sober, faithful mind replied. *And you cannot see.* But the larger part of me didn't believe it, and I felt a strange, intense curiosity about the stand. *Is it dark in there?* I wondered. *Cool? How will a breeze sound in the leaves?*

Starting through the alfalfa, I followed the line of beaten-down stems in the direction of the stand. A wall of cornstalks drew near, stretching upward with every step. *A bear . . .* I said to myself, but the thought was overcome. I stood near the planting's edge, listening and watching with every fiber of my being. Where the trail entered the stand, a clump of flattened, blanched stalks protruded like a doormat or a tongue.

Peering through a narrow opening in that wall of growth, straining my eyes against shadows, I could see that the path widened into a small clearing. In that space, a circle eight feet in diameter, the cornstalks had all been forced to the ground. They lay in whorls, and it seemed that the bear had worked methodically, leaving no standing foliage and precious few ears visible among the leaves and stems. A faint smell emerged, a scent like cut hay and simple flowers.

Strangest, though, was the light. With the sun bright in the west and few clouds in the sky, rays caught the leaves and spread a soft green glow throughout the stand. Wanting to feel that cool light, I stuck out my suntanned hands. As if I were an orchestra conductor marking time, a noise began.

Leaves rattled, and stems shook as though a whirlwind was blowing through. Heart leaping into my throat, I jumped backward and fumbled for my bear spray. Wrestling the canister off of my belt, expecting a grizzly to emerge at any moment, I glanced down to slip the safety.

Three things happened then, one following so closely on the heels of the next that I was hard-pressed to order them: I looked down at my outstretched hand and saw with dismay that it held a plastic voltmeter rather than a can of pepper spray; a pair of ring-necked pheasants burst from the corn and flew off gabbling; and I heard the words "very near" and "you cannot see."

Walking out shaken, I reminded myself that many things could have flushed that pair of birds. Just my standing there could have put them up, but all the same, I kept well away from the corn. Reaching the truck and climbing into the oven-hot cab, I took a long pull from my water bottle. The tremors in my hands kept the liquid in motion.

5

The
Edge
of the
Stand

WAKING IN Missoula as dawn streamed through my window, I wondered whether bears remained in the field. I drove to the Mission Valley as the sun rose free of the mountains. Coming to the corn, I watched pale light catch the tassels, then the leaves. Falling on bare dirt between the stand and fence, it revealed a line of deep fresh tracks.

The prints began near the northwest corner and continued for ten yards before disappearing into the stalks. A smaller but equally recent set of tracks, made by a sow or yearling, were visible on the rubber mat at the outermost pivot crossing. Studying them, I was transfixed by the way all four feet had bunched together just outside the wires. Inside the fence, almost dead center on the mat's black rectangle, were two more prints, one hind foot and one forefoot, aiming straight at the corn. The tracks looked unhurried, careful, and not at all like the trail of an animal that had been shocked.

My voltmeter registered around six kilovolts as it had the day before. Though a respectable charge, it fell short of the nine-thousand-volt readings of late July. Keeping a wary eye on the nearest cornstalks, I tried to put the pieces together in my mind.

When the pin finally dropped, I could see as clear as day how it must have happened. The bear would have approached the fence warily; sniffed at the strange rubber; inspected, but not touched, the uppermost wire. She would have drawn herself up to look into the field before settling down in a compact crouch with her front and hind paws grouped together.

Moving that way would have put her entirely on the mat, and if she had gotten all four feet onto rubber before touching a wire, she would have been insulated. A bear could cross that way without getting a shock no matter how high I managed to push the wires' voltage. Kept out of contact with the ground, a grizzly could touch the fence and stay as safe as a bird on a power line.

Wondering whether the bears were smart or lucky, I walked to the second pivot crossing. The mats were again blotted, but the tracks were all outside the wires. I could see the marks of two large front feet, then a dragging, featureless smudge turning away from the field. Here, it seemed as though the bear's hind legs had remained firmly in contact with the ground, and the fence had shocked him.

Now following a grizzly trail rather than the exact route of the fence, I found a bewildering quantity of sign. Heaps of shit lay every ten paces or so in the beaten-down grass, and every muddy place held a track. All of the sign seemed fresh, with much of it looking new enough to have been made the previous night.

I made a full lap of the field, a two-mile circuit that left me feeling as if only sheer luck had kept me from tripping over a bear. Every swale betrayed the recent passage of grizzlies. Each run and segment of the fence had been tested, but it was only on the north side, at the pivot crossings, that I could see evidence of bears getting in.

I understood then that I would need to fix each of the six crossings and that it would have to be done without shutting the fence down overnight. So long as I worked at the height of day, when most bears were bedded, I could be relatively sure that none of them would test the fence. If the wires went dead in the darkness, even for one night, all bets were off.

Aiming to stop the bears without dismantling the fence entirely, I decided to expand on what Steve had done at the sixth crossing by protecting each of the pivot crossings with a pair of electrified mats. I intended, though, to come up with a better way of making them—a design that didn't leak voltage. It was only at the end of my evening drive home to Missoula, when I passed a corner where a man was unloading pallets of beer and soda from a delivery van, that a solution came to mind.

THE FOLLOWING morning, I drove my truck out to Missoula's Reserve Street, where megastores regard one another across asphalt. I cruised the unholy service roads past miles of windowless walls, loading docks, bales of crushed cardboard, old food-grade drums, and other desolate piles.

Pulling up to the cinder-block back end of a huge grocery store, I stopped beside a pile of black plastic pallets. Measuring four feet square, they were heavy, stackable, and flat topped, with extruded bits sticking out beneath that gave them a passing resemblance to muffin tins.

Walking to a small door beside a closed dock, I pressed a button marked DELIVERIES and heard a buzzer ring within. A shuffling sound reached me, and the door swung open to reveal a small older man blinking in the morning sun.

He was thin and stooped, his face sallow and deeply lined. Deciding that I was not a deliveryman, vendor, or garbage-truck driver, he straightened, hooked his thumbs behind his soiled apron, and waited.

"Good morning," I said.

He looked up and down the street. Seeing nothing but my pickup and the usual commercial wrack, he asked what he could do for me.

I explained the situation with the bears and the corn, how the fence had failed, and how I was trying to fix it. As I talked about the grizzlies, his eyes grew intent behind wire-rimmed glasses.

"Here's the thing," I told him. "To keep bears from crossing the fence, I need to build more electrified mats, and I figure that the best thing to make them out of would be a dozen old soft-drink pallets.

"Like those there," I said, pointing to the stack thirty yards down the wall.

"Oh," he said, losing the hawkish look. He shook his head. "I can't. Wish I could, but we don't give anything away back here. Nothin' at all."

Retreating by a step into the loading dock's gloom, he moved to shut the door.

Meaning to stall him, I asked what was usually done with the pallets.

He paused, looking embarrassed.

"Distributor's driver's supposed to take them. Supposed to, but sometimes he doesn't, and they sit. That's how come we've got the pile."

Shooting a glance over his shoulder, he dropped into a whisper.

"Sixteen bears in that field, you said? God, that's a lot. I grew up in the Bitterroot, and we had bears. Black, not grizzlies, but still."

He peered around the doorframe at the pallets.

"How about this?" he said. "That driver won't miss a few. No, he wouldn't."

Looking me square in the face, he said, "I'm going inside, and if that stack's gone when I come back—well, it won't bother me a bit."

I started to thank him, but he was already turning back to the loading dock's dim interior, closing the door.

"Sixteen bears," he said as he went. "I'll be goddamned."

Loading the truck with a fine sense of truancy, I nestled pallets into the bed and lashed the stack down with ratchet straps. I made one more stop at a farm-supply store for a roll of four-foot-tall woven-wire mesh, brought the materials home, unloaded them in front of my garage, and went to work.

Over the course of a long afternoon, I fastened a section of the sheep fence to each pallet. I ended up with twelve unwelcome mats built to my new design, each covered with a grid of electrifiable wire. So far as I knew, nobody had used a shipping pallet as an electric-fence insulator before, and I was pleased to be breaking ground in the field of bear deterrence.

The prototypes were promising when I tested them in the yard. They didn't leak voltage, allowed the wire mesh to rest several inches above the ground, and wouldn't drain a fence's strength. Light and portable, they would be easy for a single person to install.

THE FOLLOWING day was a whirlwind of activity. Arriving early at the field, I set pallets on either side of the pivot track at each crossing and pinned them in place by driving spikes into the ground. Stringing strands of high-tensile wire between each pair of unwelcome mats, I built a rudimentary hammock that hung just a few inches off the ground. I was careful to leave enough slack for the pivot's tire to roll over without trouble.

The results weren't as pretty as I had hoped they might be. Set on rough ground, the pallets refused to lay flat. They looked as though they might have fallen from a passing truck, but I judged that no bear could cross without touching a wire.

That afternoon, Brittani Johnson, the graduate student whose Blackfoot Valley study had sparked my interest in fencing Schock's corn, joined me at the field. She wanted to see the results of my work and had brought along a half-dozen motion-sensitive trail cameras. We installed these on posts, training some on the crossings, hoping to learn whether the new mats would be effective. We turned the others toward the corn to find out how many bears were in the field. The cameras shot regular video through the day, shifting to infrared with the onset of dark.

As we switched the cameras on, a strong feeling gripped me —a consuming desire to witness and encounter. Rigging the last camera among the outermost stalks, I wanted to meet a bear. The impulse felt dangerous, and I did not speak of it to Brittani at the time or to anyone afterward.

The last thing I did that day was tighten the fence's top wire near the innermost sprinkler crossing. It was a minor adjustment, a single turn of an in-line strainer. I made the change for form as much as function because the wires bowed slightly between posts.

I put enough tension on the top wire to get it taut and trim. Pleased with the results, I turned the fence back on. Checking the voltage left me feeling pleased. With the new mats connected, the fence climbed back above seven thousand volts.

ONE GRIZZLY is enough to change a place. A single encounter can imbue a grove or a thicket with a numinous quality that remains far longer than the animal itself. One is enough, but near Millie's Woods, there are many. Any low or sheltered place, any cattail marsh or patch of brushy hawthorn trees, may hide a bear.

August was giving way to September, and distant forest fires hazed the night. Stars showed dimly, the wind was halfhearted, and noises seemed muffled. From time to time, sows and cubs spoke in their guttural language: a plosive cough of warning, a grunt or something like a deep thrumming snore to prove that all is well.

They moved in the night, breathing deeply. Wide paws fell soundlessly, and damp soil — their preferred spots are often moist, as ours would be if we went barefoot — deformed beneath their weight.

On such a night, Schock's pivot sprinkler rolled south across my electric fence, and some bolt or protuberance on it caught fast to the taut upper wire. Nobody was there to watch the tires creep ahead or hear the tower motors groan. Grizzlies were nearby, though, making their rounds of the valley floor.

Perhaps the nearest one heard the wire's complaint, a noise like a guitar string tightened by a careless hand. When the strand parted and thirty yards of high-tensile steel rippled through the grass like a mad snake, tangling, striking at tires and uprights as the pivot trundled into the corn, any bear in the area would have stopped to listen.

Night wore on. The pivot passed into the stand, and it was lucky that the wire did not curl too tightly around either axle, bending instead into loose snarls that trailed behind the tower wheels like silver hairs. Hour by hour, the sprinkler worked deeper into the field.

MORNING BRIGHTENED, and I came up from Missoula. Finding the fence torn, I hunted for the missing top wire in the open, canvassing the alfalfa and searching the track muck. I looked seventy yards across the corn to where the sprinkler was. Its noise came clearly across the distance, a constant chatter of droplets against leaves.

I did not enter the stand lightly or carelessly. Despite my yearning to see grizzlies, I was afraid. I remembered Greg Schock's story about driving out of the field and mowing down his precious crop to keep from being on foot. He had told me not to walk in the corn when ears were on it, and I had believed him.

But just then, three things were clear to me: First, the pivot was still functioning properly. Second, I did not want to tell Greg that I had entangled an expensive piece of equipment unless I had to. Third, it was full light and the bears, wherever they were, had likely settled down to rest for the day. On top of all that, I wanted a reason to enter the corn. After all my time spent at the stand's edge, I needed to see what it was like inside.

There had been no rain for a week, but the sprinkler's passing had turned the dirt to a gumbo that caked my boots. Stepping over my fence, I crossed the field's bare margin. Tick waited behind the wire, whining as I drew farther away. Marshaling his courage, he squeezed beneath the wires and came to my side. We paused together where the wheel rut met the stand's weedy edge. An old pat of bear shit lay at our feet, but Tick was too familiar with the stuff to sniff at it for long. He stayed close as I unfastened the bear spray's Velcro keeper.

With one hand on the spray and the other clasping a set of wire nippers, I entered. It was hard going. With no perennial roots to hold the soil together, the tire ruts were deeper within the corn than outside of it. In places, the bottom was a foot down and made of sucking mud.

I took a step and nearly lost a boot. Another, and I stuck fast. Struggling only sank me deeper in the morass. Tick watched. With nothing to lean on or pull against, I had to bend down and use both hands to break free. I stood up muddy and gasping, wiped my palms on the nearest plants, and afterward straddled the rut.

Corn grew close to the track, coming to its brink on one side and leaving a few inches of bare dirt on the other where the hub and overhanging motor traveled. Tick could walk that narrow path, but I had to hobble bowlegged with a boot on either side of the rut. From time to time, my left foot, which was on the narrow side, slipped in.

It was darker in the corn. The morning sun was still low, its light strained through leaves, tassels, and silk.

Everything was green and gold, and after every few steps, I paused, waited, and listened. I looked ahead ten yards to where the wheel's path was hidden by its curve, then glanced back the way I had come. Peering sideways into the growth, I found the leaves and shadows as impenetrable as a wall. Saying a word to Tick, who stayed mercifully at heel, I pressed on.

I went deeper. It was a good while before it struck me that I ought to announce myself. Though I knew that it is best to be loud when bears are around, I was loath to break the corn's silence. With its stillness and green stained glass light, the place felt like a chapel. Shouting seemed like sacrilege, as likely to anger bears as scatter them.

I settled for low and constant singing, maintaining a bow wave of sound without regard to content or pacing.

"In spite of ourselves, we'll end up sittin' on a rainbow. Against all odds," I sang as I crept through the corn. "Honey, we're the big door prize." I paused, waited, and glanced behind me. Tick watched, concerned.

I reprised that John Prine song many times over. The dog stayed near, the going was slow, and I hummed when I ran out of words. With every synapse occupied by nervous observation of the surrounding corn, I stood on the wire's trailing end before I saw it.

A few more steps—now through standing puddles—and I could

look down the wheel track's narrow slot to see galvanized pipe, tires, and nozzles throwing spray in white arcs. Water rattled on leaves, and from time to time, the sprinkler's motors kicked on with low groans.

The machine's presence was a comfort. It broke the silence and did away with some of the field's strangeness. A pivot tower could be climbed in a pinch, too, which reassured me as I neared the outermost reach of the nozzles.

I worked quickly, slipping in and out of the spray, snipping coils away from the axles. By some lucky chance, nothing had been bent or broken.

I had to be careful when the motors stirred, lurching the belly-high tires forward. The wheels had heavy tread, and I had heard stories of farmers caught and crushed by their slow, implacable motion.

When possible, I stood away from the sprinkler as I cut it free, pecking at problems like a fishing heron, dodging in and out while water poured down and soaked my shirt.

In time, I reached an impasse. Between the tires, wire had snarled into a mess that could not be unwound without crouching in the path of the oncoming wheel.

Standing to the side, I heard the motors kick on and saw the tires slide forward by a foot. I counted to forty before the machine stirred again, then dodged into the heart of the spray, blinking water from my eyes and cutting for all I was worth. I tried to keep track of the seconds, losing count around twenty. Making a final effort with the pliers, I felt the bunched strands slip free, and I leaped back as the sprinkler rolled ahead.

Retreating until I was no longer being watered, I fashioned a coil from all the bits and pieces of wire and set it by the rut. I noticed that the dog was missing and that the cornstalks to my east glowed with brighter light.

"Tick!" I called. His name rang scared and harsh in my ears. "Come!"

I heard nothing for a moment, then the sound of jostled leaves. Bursting into view as an actor ducks through curtains, the dog trotted to me.

Crossing to where he'd emerged, I saw a patch of sunlight through a thick screen of growth. The light did not seem far off. Parting stalks with my hands, I left the rut.

Walking the pivot's route had been difficult, but traveling through the untracked corn was nearly impossible. Plants grew ten or twelve inches apart, and they were as hard as broom handles. The leaves — some of them now edged with brown — chafed my neck and forearms, leaving marks that stung like paper cuts.

The dog made easier work of it. Short enough to clear the lower leaves, Tick slipped deftly along. Bending low, I tried to follow his example. Stooping helped, but it forced me to stare at the ground. Moving blindly, seeing only soil and roots, I shuffled on.

I was about to turn back when the leaves split to reveal a small clearing. The ground was paved with trodden stalks, and sunshine gleamed on half-eaten cobs. On the far side of the space, a three-foot-wide hollow had been scraped into the soil. Though the pivot had soaked the spot, sprays of fresh dirt still patterned the ground. Two smaller excavations had been made at equal distances from the larger digging, and their look brought to mind a planet orbited by moons.

With a growing sense of disquiet, I supposed that I was standing over the daybeds of a sow and her cubs. Walking to the center of the opening, I took a knee and plucked up a gnawed cob. Its kernels were plump and yellow. Pressed with a thumb, they exploded in gouts of juice.

Months ago, Greg Schock had told me that grizzlies ate corn in the same way that a person would, by grasping with a paw and chewing along the side. The half-consumed ear hinted at a delicate touch. Human teeth and hands could have done no better.

Across the clearing was a small interruption in the green pickets, a gap made by knocking down two rows of plants. Calling Tick, I went to it, looked through, and entered a labyrinth shaped by the hunger of bears.

There were rooms and galleries, halls and bedchambers. Feeding grounds, some of them ten yards across, were connected by corridors and anterooms. The light was always green, a wind kept the tassels in easy motion, and I moved across a carpet of flattened stalks. Wonder and terror wrestled in me while I walked, and when I dropped to a knee to examine some daybed or half-consumed cob, the leaves above me reached toward one another, giving the impression of an arched ceiling — a nave with sky at its center.

Passing through the places where grizzlies fed and slept, I moved deeper into the field's silence. Path led to path, opening to opening, and the sky above was blank and blue. It was cool and damp, with the ground soft from irrigation and broken stems exhaling their cut-grass smell. There was something captivating, even intoxicating, about being there. The dog did not like it much, but to me, it felt like swimming beneath the surface of the ocean. For a little while, there was only peace, quiet, and the rhythm of walking on.

I floated that way, with my mind at some remove from my body, until I arrived at the center of a wider clearing. The sun had climbed higher and now shone fiercely on wet leaves. I squinted against its glare.

I remembered that the day was passing and that work awaited me on the fence line. Looking back the way I had come, I saw three trails departing from the clearing's far side, each striking off in a different direction through the corn.

The sun worked against me, flashing leaves, pouring shadows on the ground. I thought of the old game-show staple of having a contestant choose between three closed doors. Hoping not to hit the jackpot of a furious sow with cubs, I cast my lot with the likeliest path. As I walked it, the corn looked so unfamiliar that I nearly turned back. In the end, I kept on, knowing from the sun's position that my heading was more or less correct.

I traveled too fast, making split-second decisions when turns loomed and ways diverged. The ground was flat and featureless, and it struck me that I might have to choose a straight path out, blundering on a blind course while stems rasped me raw. Fear burned higher at that thought, my quickening breath stoking it like a bellows.

Hoping not to make things worse by hurrying, I shortened my steps and soon began to recognize spaces that I had passed through. Hearing the sprinkler, I navigated toward it.

Pressing back through the curtain of standing corn, I met the muddy, even curve of the wheel rut. Gathering the nippers and coil of wire, I started out. It seemed a long time before a slice of the outside world yawed into view.

Getting clear, I turned to see that the stand had snapped shut like a leg-hold trap. From where I stood blinking in the sun, the dark slash of the pivot track was barely visible.

MY GRANDMOTHER, reading this, will take me to task. She will ask why on earth I chose to walk in the corn that day and, in particular, why I went on farther than was necessary to cut the wire free of the sprinkler's wheels. She will threaten to stop praying for me if I don't quit taking such risks, though I don't think she'd ever follow through. The reason that I'd give her, if she pressed, is that I wanted to see what the bears had wrought and understand how they were living in the corn. I would tell her that I was, in a word, curious.

But curiosity, whether professional or personal, was not sufficient rationale for pressing on through the first or second clearings that the grizzlies had made. Neither does it express the peace and pleasure that I got from being in the stand or the sense that I had of being drawn irresistibly on.

Something old, mysterious, and real took me in among the overlapping leaves. Calling it bear madness would not be too strong; there is something about grizzlies that fascinates and pulls at everyone who encounters them. Half of it is their odd mix of bulk and grace. A healthy bear is a walking paradox — a heavy, seemingly ungainly creature that can turn swift and lithe in a moment. Near Yellowstone, I have seen a boar running faster than a sprinting horse can, stretching out and loping across an open meadow. Coming to a stand of brush, that bear stopped short. Losing all evidence of strength and speed, he went lumpish and shambled from sight. The phase shift was as complete and unlikely as water flashing to steam, and I never forgot it.

Then there are the eyes, which are not unlike ours in shape, size, and distance from each other. A bear's eyes — small in their wide heads — seem made for looking back, for focusing on and assessing us. When grizzlies are not afraid or raging, there is something tranquil, sympathetic, and even shy about their eyes.

When a human meets a bear, their gazes join like halves of a split stone. A charged arc is struck between two creatures, and the rest of the world disappears in the glare. That fire is treacherous and tends toward destruction. It also contains a measure of recognition.

Nobody can say what a grizzly makes of that moment. Most bears, on most days, seem to regard a human being as a minor nuisance or a hazard to be avoided. Seeing a person, they recede into the nearest heavy cover. Others, though, consider a human to be either an intolerable threat or a morsel, and they act accordingly.

I do know what the experience means to me. Looking bears in the face—an experience as consuming as falling—has given me a better grasp of what I am and how I fit into a wild older world.

Perhaps I was courting that feeling when I went into the corn, though I would never have said so at the time. I wanted to draw near enough to be seen, known, and spared by a grizzly. The bears, fortunately, were less eager to encounter me.

MILLIE WAS busy in August. She had apples—all that she could eat—but the corn was fenced and shut to her. In another era, she would have gone to higher country. Feeling the end of the season, she might have made for the groves of whitebark pine that grew on the peaks to raid caches made by squirrels. Eating oil-rich pine nuts by the bushel would have put a final polish on her preparations for winter.

But many of the whitebarks were gone. Pine beetles had girdled and killed the trees, turning whole slopes from green to red. Winter's deepest cold should have checked the work of the bugs, but it had not come for years. Only a true killing freeze reaches beetle larvae in the cambium where they take shelter during the dark months. An early cold snap is best, with the temperature plunging below zero before the insects can manufacture the natural antifreeze that sees them through to spring. For too long, no cold like that has swept down from the north into Montana.

Instead of climbing into the mountains, she stayed on the valley floor. Millie and her cubs must have come near the field in those days. She knew the landscape well and had foraged in it for years. Stealing down from the woods, she would have encountered the fence. I can picture her nosing against a wire, feeling the shock, and sprinting away.

Beetles and warmer autumns had taken the pine nuts. My fence surrounded the corn. But bears are creative when it comes to feeding. If there is a meal to be had, they will eat.

She took her daughters north toward Post Creek. Cautiously, they dug gophers from the alfalfa fields at night. Her cubs worked avidly, tossing up dirt and swallowing all the peeping, fleeing creatures they found. They hunted that way until a blue glow appeared over McDonald Peak.

Millie saw the dawn coming. She knew that in the Mission Valley daytime belongs to human beings as surely as night does to bears. When the window lights of nearby houses snapped on like wide eyes opening, she led her cubs into the sheltering woods.

But something on the wind kept her from bedding in the grove. Lifting her nose, Millie peered south through aspens. Catching the sharp note of rotting meat, she led her cubs on.

Grizzly bears are very practical where food is concerned, and though the stink wafting through the trees would not have brought a human running, Millie knew what it meant. The smell of rancid flesh is unmistakable. It is rich with the promise of fat, and fat keeps a bear's heart beating through winter.

She gambled, bringing her cubs through the brightening world toward a house on the edge of the woods. There she found a garden of ursine delights, with elk hides hung out to dry and overstuffed garbage cans. Millie turned to the work of feeding, eating all that she could.

Engrossed, she did not notice the house's screen door opening. She did not see the long barrel of a twelve-gauge shotgun in the dim light. The sound of a shell being chambered, if it reached her, came too late. Morning shattered at the seams, and for an instant, Millie saw white light as if the sun had jumped above the mountains. Lead pellets struck her head-on, shattering the bridge of her nose and burying themselves in the skin around her eyes. In their wake came darkness and scalding pain. Tasting blood and seeing almost nothing, she ran blindly with the cubs at her heels.

6

Seeing

THE TRAIL cameras sat on the edges of the cornfield, each screwed to a freshly driven post, staring glassy eyed at weeds and stalks. Turned on and off by motion-activated sensors, the cameras kept watch day and night. Through the sunlit hours, they recorded high-resolution, full-color video. Filming by infrared light in the dark, they captured ghostly black-and-white footage that rendered common things strange.

A cold night came with the start of September, and morning revealed a transformed world. Clumps of grass glinted white in the sun, and when the rime burned away from seed heads and stalks, it left them changed. The shift happened quickly and was most striking in the corn where frost parched the leaves. At sunset, the cornfield was a green oasis. By morning, it was tinged with sepia, with leaves drooping heavily and ears cracking yellow-toothed smiles.

Four of the trail cameras stood at pivot crossings, seeing only wind-jostled grass and occasional deer moving along the cornfield fence. The other camera was inside the perimeter of my fence, facing out from the edge of the corn, in a spot where a skipped row left a narrow path of bare ground leading into the stand. The place turned out to be a crossroads. The first time that I retrieved the

camera's memory card, coming home late from the field and settling down on the couch to watch the videos with Gillian, I found a
record of busy wildlife.

The daytime footage was nothing special, with trucks passing
on the road and birds speeding arrow-swift through the frame. But
night brought whitetail bucks in tattered velvet and does feeding
close beneath the camera, their heads falling slowly to graze, then
rising alert. One deer sniffed the camera post, setting the frame
shaking and spooking her fawns. Flocks of Hungarian partridge
marched back and forth, taking short neat steps as if they were
troops on parade. Sometime after that, a lean fox wandered from
the corn, paused to scratch itself in the dawn's glow, then slipped
off toward the mountains.

We worked through the files on the card, amazed by the variety of life and the easy way that animals moved through the field.
Each video was time stamped, showing with certainty that pheasants had come through at 2:15 a.m., a fox at 3:54, deer at 4:00, then,
incongruously, a fat tabby hunting mice as the darkness waned into
day. All the images were gray scale, with the eyes of each mammal
flaring palely in the camera's infrared glare.

A weekend's worth of recording was on the card — fifty-some
minute-long videos — and we watched the first day's footage
straight through. Then, because our own night was getting on, we
began to skip through files faster, clicking ahead once we saw a
deer's hindquarters or the trailing tail of a pheasant. We dismissed
each video in turn, saying: "Deer . . . nothing . . . nothing . . . partridge . . . wind . . . nothing."

I had grown so used to seeing delicate creatures that I hardly
recognized the bear. The footage had been shot at night with the
infrared flash, and before I set the video playing, I did not register
the black shape blocking most of the frame as an animal. I thought

that dirt or a windblown leaf had covered half the camera lens. Then the thing began to move, resolving into a shoulder.

The grizzly was close, with its neck nearly at the base of the post that held the camera. Frozen where we sat, Gillian and I watched it step back and lift its head until a white-lit eye peered straight into the lens. After glancing toward Hillside Road, then into the corn, so we saw one side of its face, then the other, the bear turned broadside.

Though I sat safely in my living room, fear gripped me. There was a window opposite the couch, and I grew conscious of night pressing against its pane. The bear was huge in profile, with a shaggy hump rising from the line of its back. Its forelegs were massive and dark, their fur in perfect order. Small round ears set high above a broad forehead leaned forward with eager intent. Below those ears, the bear's muzzle looked as solid as an anvil. Behind them, its neck was as thick as my chest.

The grizzly looked the very picture of health and must have weighed five hundred pounds. Stepping lightly for so large a creature, it passed through the frame — black nose, dished face, and high shoulders giving way to a backswept line of blonder fur and the rump's smooth curve. The bear accelerated as it went, its glossy coat shivering, rippling, and bunching with accumulated fat. It gained speed with every step until massive queerly human legs propelled it from the frame. Passing through scattered low pigweed, it left plants shaking in its wake.

We watched over and over, at first speechless, then remarking on the deft urgency in the grizzly's movements. As the footage played on loop, we discussed whether the bear had been driven off by some smell or if the camera had made a noise when it switched on.

Clicking through other videos — more deer, a waddling skunk, and the incongruous tabby making for a farmhouse's distant light —

we watched the minute-long highlights of a Mission Valley Sunday night. We were nearly through to dawn, with the time stamp reading 4:45 a.m., before a second bear appeared on-screen. Caught in the act of entering the frame, the grizzly stopped short. It showed black against the pale soil of the field's unplanted margin, and after taking stock of the night, the bear swung its head, looking like a turreted cannon traversing a battlefield. Facing the camera, it fixed us with a grave unblinking stare before turning along its length to vanish from view.

I SLEPT poorly that night. The bears gave me no peace. Tossing between the sheets, I dreamed of grass beaten down as if cattle had come through, with trails bending among gentle hills. Tick was with me, and together we descended toward a house in a wide dell.

The place had the grandeur of a fallen plantation and many broken windows. White paint had surrendered to time and weather, the boards beneath showing like bone through torn skin. The walls had split and collapsed in places, leaving holes that a man could walk through without stooping.

Following the dog through one such gap, I found myself in a room with high shelves and wide windows. Looking through those windows at the latticework of trails in the grass, I saw the bear. It did not crest a hill or round a bend but incorporated like a cloud on the brow of a mountain, shimmering into being.

It was faint, then definite, then suddenly running. Loping a straight line, the bear made for the house. He was small in the distance, but I could see him stretching out, becoming pure muscle, arriving like a storm.

Half a breath and he was in the room, terribly large among couches and chairs. Tick ran like mad, and the grizzly gave chase, snapping at his heels.

The bear's efforts broke the room apart, shaking the house until plaster showered from the ceiling. Tick ran on but was tiring. He came to bay under an old coffee table, with the bear huffing and slavering close beside.

When I bolted for an open door in the far wall, the grizzly turned, lunged, and dragged me to a stop. Pressure crept up my arm, through my shoulder, toward my heart.

Twisting free, I fumbled bear spray from my belt and shot the whole works into the creature's pink-gummed maw. Shaking and bellowing, the grizzly turned away. I ran, slipping through the doorway with the dog at my heels.

Slamming the door home, I looked up to find that I could still see the bear through a pane of cracked glass. On the other side, the creature licked and snorted its nose clean of pepper spray. The small dark eyes stayed fixed on me.

I WOKE early and restless on Saturday morning, roused Gillian, and loaded our Subaru with bedrolls, hiking boots, and all we'd need for a weekend in the mountains near Glacier National Park. Atop that pile went my voltmeter, an Altoids box full of memory cards, and a large shockproof plastic case containing a brand-new quadrotor drone.

I have never been a natural twenty-first-century man, and technological advancement comes hard for me. Because of this, I received the news that People and Carnivores had purchased the drone with mixed feelings. I disliked the thought of harrying the bears and doubted that I could learn to drive the thing.

Still, the prospect of seeing into the corn captivated me. Prior to the drone's arrival, I had entertained various theories about how best to peer inside the stand. Even the most feasible and reasonable of those notions — rigging a harness to the pivot sprinkler and riding it through a twelve-hour circuit — struck me as the sort of thing that would sit poorly with People and Carnivores' legal counsel.

By the time Gillian and I headed north from Missoula in the car, I was starting to feel more like a technological convert. I watched the land pass with mounting excitement and soon caught sight of the Missions reaching toward a clear sky. Hands clumsy with nervous energy, I unloaded the plastic case onto the gravel shoulder of Hillside Road, spun four plastic propellers onto driveshafts, and powered up the machine.

The drone took off in a flurry of dust, whining and rising until it was a dot in the blue. I watched the screen of a tablet connected to the controls, looking down at the world through the machine's camera.

The thing was wonderfully stable, and it hung steady at two hundred feet while I panned the lens downward. The car's small rectangle came into frame as did the dots of our heads.

Beyond us, the electric fence was visible as a silver thread, and the field's dirt apron looked like a tawny beach beside an ocean of growth. Toggling the controls with my thumbs, I sent the drone to sea. It raced off with a yellow jacket's speed until I could no longer hear the motors.

I saw the field as never before. From above, the corn was a wide swath of uniform growth bull's-eyed by wheel ruts. In places, the crop was marred with dark blotches where bears had slept and fed. I descended until I could see rows in straight-lined perfection and the crumpled stalks lying in the bigger clearings. Flying lower still, I expected to see a grizzly rise from a daybed and run.

Hoping to see movement, I investigated one trampled spot, then another. Gillian watched the screen with me, inhaling sharply when I flicked the wrong joystick and nearly crashed. Worried that

I might lose the contraption on its first day out, I settled for a bird's-eye view, flying high and trying to get a sense of where and how much the bears were feeding.

What I saw was heartening: Though clearings existed in the corn, they were scattered, small, and confined mostly to the field's northwest corner. They looked like the work of one or a few animals, and the damage fell far short of a 20 percent loss — the figure that Greg had named in seasons past.

Near the middle of the field, in line with the center point of the pivot, bears had been feeding in a straight row. Elsewhere, the clearings were scattered, but here they ran along a north-south axis. I could not make sense of the pattern until the drone cleared the edge of the field and whined earthward, and I remembered something that Greg had told me when first we talked about the bears.

"They're smart," he had said. "Even picky. We planted a few acres of sweet corn, and they ate that down to nothing before touching anything else. I guess that's no big deal — you or I would do the same thing. But then I noticed how they ate in the big field. It's all the same seed out there — all feed corn — but it's not the same to them. There's a spot where old property lines come together. The far side used to belong to the neighbors, and the old-timers plowed with moldboards, which threw all the dirt to one side. The soil along the old boundary ended up thicker over the years, and the bears know it. They always feed in there, looking for the best, ripest stuff."

I was impressed that bears could be so discriminating and thought about it often as Gillian and I passed the weekend in the mountains. Coming down from a clear, high lake, we found ourselves waist-deep in huckleberries. Stopping to gorge on purple-black fruit, we left our packs and foraged the overgrown hillsides. The berries were perfectly ripe, and their endlessness drove us into gleeful frenzy. I stripped fruit from branches with both hands, heedless of all but accumulation.

Staying long in the patch, we filled our stomachs, water bottles, and every other container we'd brought that could be trusted to make the trip home. When we finally left the peaks behind and headed south for Missoula, afternoon was well advanced. By the time we stopped at Schock's dairy, dusk was building above the Mission Range.

Stepping from the car with bear spray on my hip and the tin of fresh memory cards in hand, I shut off the fence's current and climbed into the field. The trail camera remained where I had left it, watching over weedy, hard-packed earth. I switched the power off, removed its memory card, and put a fresh one in the slot. Walking down the line with one eye on the corn, I repeated the procedure at each of the other cameras. As I slid the final card home, the phone rang in my pocket.

"I see your car," Greg said. "Come up to the house if you want to see some bears."

IN HIS ample living room, across from a large glassed-in fireplace, Greg bent over a spotting scope on a tripod. The scope was positioned to look out the room's wide windows across the eastern pastures toward Millie's Woods and the mountains.

"Had a milk cow wear out a couple days ago, and since I put her in the dead pit, they can't leave her alone. Got her pretty well cleaned up already," he said, straightening and making an open-handed gesture toward the scope.

Bending to the reticule, I saw nothing but grass.

"Swing it up," Greg said. "The pit's on top of the dike by the pile of old concrete and pipe."

While I walked the scope's field of view up a raw-soiled berm, Greg held forth to Gillian.

"Corn's coming pretty good this year. This time last summer, it looked rough. Patchy. But then again, it was a different kind of year."

I swiveled the scope toward where a long, low hump of dirt created its own horizon. Seeing movement on the rim, I dialed up the magnification.

"I mean it was dry. This year we've had rain. Not a lot, but we've had it when we needed it. No late frost, and we've had heat when we needed that, too. You look at the apple trees. Look at the fruit, and you can see the kind of year it's been."

He went on, but I heard nothing more. Twisting a ring, I brought the scene into focus. In exquisite detail, I saw a cleared space punctuated with bent irrigation pipe and other agricultural debris. At its center, a sow and two cubs worked feverishly at the task of consumption. They bent low over the carcass, and even from a distance, I could see how they wrenched flesh from bone.

The sow was an avid feeder. She dropped her muzzle, took hold of some savory bit, and yanked with front legs braced, like a dog playing tug-of-war. Sometimes the carcass slid and her cubs stood away, looking to her for a moment before getting back to business. From a distance, with their heads low and their back lines bristled with humped-up fur, the grizzlies brought to mind sea urchins moving on the ocean floor.

Taking turns, Gillian and I watched until the light failed. The sow and cubs kept at their task, and I did not tire of looking. They were still eating when dusk fell. As I walked to the car, it seemed good and frightening to be outdoors with them in the gathering dark.

MILLIE SHOULD have been feeding the same way and teaching her cubs to survive in a complex world. She ought to have been ranging widely with them across the foothills or climbing high to raid whitebark caches and army cutworm colonies on McDonald Peak. In past years, she would have been growing sleek on the produce of late summer. The cubs would have been fattening, too, growing larger and stronger by the day. The three of them would have walked together through the freshening night.

A contented grizzly cub makes a noise like a kitten purring, only deeper. Millie should have heard it on those evenings and felt what a sow is capable of feeling when surrounded by her healthy, growing young.

Instead, grievously hurt, she blundered through the hayfields and forest near Schock's dairy. She kept to the heart of her range, visiting the places that were most familiar to her.

Her eyes, particularly her left one, swelled shut or nearly so. The bridge of her nose had not healed, and in places, small holes were punched through the pale exposed bone of her septum. She could barely see, and the magnificent organ behind her nostrils was irreparably shattered. She could not smell a thing or even pass air through her nose.

Robbed of its nose, a bear is adrift in an inscrutable and treacherous world. Millie felt her way from place to place, seeking food as a grizzly must at the end of summer. Without scent or much sight to guide her, she was more often hungry than full. When she did happen across something—a road-killed deer or garbage left out overnight—Millie ate with a desperate will. The very act of feeding, though she could not have known or avoided it, was part of her undoing.

Infection is hard on grizzlies. Though the animals are as tough as nails and capable of withstanding immense physical strain, bacteria can bring them to ruin. Years ago, Stacy and other biologists on the reservation noticed an alarming trend: Grizzlies trapped for research and relocation were being found dead weeks or months after being released. In time, they learned the simple, unfortunate cause. Frightened confined bears often ended up smeared with their own feces, and a hypodermic injection — necessary for anesthesia — given through such contamination introduced enough bacteria to lay a grizzly low.

The same thing holds true for other wounds and sources of infection. Because grizzlies, being scavengers and opportunists, seek out rotting meat, an unscabbed pinprick in the wrong place can do a lot of harm. Millie had much more than a prick, and by early September, she had been living with open sores for at least two weeks.

Nearly blind, she remained a slave to hunger, eating anything that she could find. Rotting meat, when she could get it, was spread across her muzzle by the act of feeding. She rubbed at the places that troubled her — so much so that she wore away hair and hide — smearing unwholesome foreign material into her wounds.

Her milk ran dry. The cubs went hungry. Soon afterward, Millie drove her offspring away. I can't say how that might have gone, but I know how devoted sows are to protecting and rearing, and how much a cub desires to keep near its mother. I cannot imagine that the severance is peaceful, painless, or easy to watch.

They parted ways, and Millie headed south toward Schock's cornfield. Her daughters were left to their own devices, to fend for themselves in the pastures and backyards along Post Creek.

AFTER DRIVING home from Schock's place in the dark, I settled in to review the footage from my trail cameras. Almost all the recordings were of windblown grass. It was late, and I was bone weary from two nights of hard-ground sleep. Drowsing, I clicked through all the card's nighttime videos, finding nothing. My eyes were closing of their own accord when I reached file 0005, but they shot wide open when the video began to play.

The footage, recorded just inside the field and stamped 8:26 a.m., September 11, showed a still morning under a powdery sky. The Missions were hazy and blue, the grass gone to seed, and the wires of my electric fence visible as three silver horizontal lines above the sprinkler's wheel rut. In the distance, I could see the broad arch of Greg's dairy shed. Much nearer in the foreground, inside the fence, was the high back line of a bear.

She — I did not know her name then, but from the first, I thought her to be a sow — moved obliquely off from the camera, heading toward the fence at the angle hunters call "quartering away." Walking in that direction, she passed the camera without showing her face.

Something was wrong. The sow's fur was lusterless and matted, dreadlocked with mud and thistle heads as if she had lain too long on her side. Ribs showed through her careworn hide, the bones of her hips were visible, and her spine stood out like a tent's ridgepole. Compared to the drum-tight plumpness of the other bears I had recorded in and near the corn, the sow looked like a deflated balloon.

She moved slowly forward, reaching my fence line at the pivot crossing.

When she turned her head eastward toward the mountains, I could see the ruined geometry of her face: the swollen pinkish tissue forcing her eye shut; the weeping wounds pockmarking her broad cheek; the faltering and crenulated line of her profile, with a raw keloid mass rising from what ought to have been the bridge of her nose.

I stared, trying to resolve the confusion of scar tissue and sores into a countenance. She turned away. Looking north, she swayed left and bumped that side of her head into a clump of canary grass as if she had not seen it. Recoiling from the touch, the sow centered herself once more in the pivot crossing, with her nose just short of my fence. Moving quickly for the first time, she shifted weight onto her hindquarters and lifted her shoulders as if to stand on two legs.

She'll jump, I thought, but she settled back with her nose pointing like a compass needle at the faraway trees of Post Creek before twisting west to examine the grass that had brushed her. As she turned, I could see that the left side of her face was at least as bad as the right. On that side, her eye had all but disappeared in a ring of sores. When it blinked — it did so twice — a subtle flinch clouded her features. The sow examined the grass for a moment, her tongue slipping out and dodging upward across her lips. It drew my attention to the blackness of her unharmed nostrils. Carrying no wounds, swellings, or raw scars, the pristine delicacy of her nose tip threw the surrounding damage into relief.

She moved again, bending away from the camera. Peering across the matted hair between her ears, I could see past the thin wires of my fence to the grass, timber, and sky beyond them.

Deciding something, she set off with surprising grace and speed in the direction of the corn. I hoped that she'd come around to face the camera as she left, but she never quite did. She walked off at an angle, so I last saw her in something just a bit more than profile. The fur along her brow was twisted and clotted with mud, mussed like the hair of a sleeper. Below her eyes, the hide was darkened with dried blood and running pus. Her tongue was always working, licking everywhere in a horsey way, moving busily as if to clean her many wounds.

She walked from the frame, pausing abreast of the camera before pressing on. The recorded sound of her steps faded. The rest of the video showed only the shining line of my fence and bits of orchard grass, bent but unbroken, which recovered from her passing and reached for the sky.

All tiredness gone, I watched that video over and over until I knew it by heart. I memorized the sow's cuts and sores, studying the way she had approached, then turned from the fence.

"Look at her," Gillian said once she had seen the video. "What happened to her?"

We watched again. It seemed to me that the sow had wanted to head for the mountains and the wires had stopped her. The notion that my fence could entrap a bear as well as exclude it sent a pang of guilt through me. Wondering if my summer work might be doing more harm than good, I asked Gillian if she thought the wounded bear was trying to get out.

She took the question seriously, keeping her eyes on the screen as the grizzly once more approached the wire.

"That bear goes right to the crossing," I said miserably. "She looks out."

The sow passed from the frame.

Gillian set her arm across my shoulder. "You can't know what she was trying to do. She never tested the fence or got shocked by it."

Later, I lay waiting for sleep to find me, pitying the sow. No one with a heart could have seen the video and felt otherwise. Her wounds were deeply infected and plainly starving her. In comparison with the other grizzlies that had passed my cameras, the sow was a walking skeleton.

I stared at the ceiling while Gillian's breathing grew regular and easy. The sow wouldn't last through winter, I knew. She would die in the first hard freeze or in her den if she made it there. I lay still, wondering what had happened, whether disease or violence had been her ruin, and how her final months would be.

I feared the sow, too; compromised and broken animals are dangerous. From the video, I could see that strength remained in her and that it outmatched mine. She had come to the corn to hide and feed, to heal if she could or die. Unable to move far, she would stake a claim somewhere. With at least one other bear in the field, the large healthy one that had appeared on my nighttime videos, she would be wary and fierce. Life had brought her to bay, and things would go hard for any creature stumbling into her.

Closing my eyes, I saw the sow turn from the fence and make for the corn. She passed before me, sticking out her tongue in a manner at once gentle and predatory.

7

Reaping

WHEN GRIZZLY cubs lose their mother, they lose every-thing. The milk that they have depended on is gone, and just as damaging is the loss of a sow's knowledge and instruction. Young cubs are seldom out of their mother's eye- and earshot, and their first year is one long lesson: This is the smell of a winter-killed deer; here is where fawns hide; these boulders are worth flipping; in Sep-tember, gorge on apples. Cubs need to understand the whole sea-sonal round if they are to survive alone. Half a summer will not do any more than half a wheel can roll down the road.

And then there is the problem of hibernation: At least some of the time, orphaned black bear cubs will put themselves to bed. Grizzlies, in contrast, must be shown how to den by their mothers. This fact makes solitude a death sentence for an orphaned grizzly. Even if it is lucky in finding food and continues fattening and grow-ing through fall, a cub of the year does not know what to do when the snow flies. Without a mother to lead it into the secret corners of the mountains to a den, a grizzly cub will not live to see a sec-ond spring.

In spite of their hopeless situation, solitary cubs display a strong will to live. They vocalize and search. If their mother cannot be found, they turn diligently to the work of surviving, using every strategy they possess or can devise in their developing minds.

MILLIE'S CUBS, for example, knew how to hunt gophers and find apples. They devoted themselves to these endeavors with a hungry will.

The cubs did not travel far. They kept to the pastures and thickets south of Post Creek and tried to make a living digging rodents. They did not, as their mother had long ago learned to do, anticipate the brightening sky and seek cover before sunrise.

Light came into the valley, and farmers drank coffee at kitchen tables. From windows and porches, the men and women saw two small dark shapes moving through swathed fields. Taking notice of the little bears and the absence of a sow, they began calling the tribes' Natural Resources Department.

The first reports came in, and a tribal game warden was dispatched. Parking his truck on the road, he watched the cubs dig in a field belonging to Greg Schock's father, Walt. In earlier years, that plot had been the site of the Schock family's first experiment with growing corn beside the mountains. In 2016, Walt's field had been planted with alfalfa and cut for hay. The cubs moved across the lush stubble, digging gophers as intently as any dog might and shoving their muzzles into the excavations. It made good watching, but as day ran down to dusk, the warden grew convinced that the little bears were alone. When he was sure of it, he called Stacy.

For two days afterward, Stacy couldn't find the cubs. He kept a close watch on the field and cruised all the dirt roads. His big white Ford truck became a common sight to the people living around there. Stacy drove around fifteen miles an hour and frequently stopped to study a patch of brush or shadows at the edge of timber. During a decade of working with bears, Stacy had learned to drive stealthily. I saw him often on the road, and his truck somehow blended into the landscape and snuck up on me. The machine's bulk and slow progress had a seeking, predatory feel. It reminded me of a grizzly.

Stacy kept watch on the field and the surrounding pastures. He looked in the side yards of houses and searched the nearer foothills. The cubs were nowhere to be found. It seemed they had vanished into the timber along Post Creek.

That they survived alone on Post Creek through the first days of September, when so many adult bears were ravening, proves that the cubs were both cagey and lucky. The fact that they disappeared so completely is proof that even a seven-month-old grizzly has a measure of its species' talent for keeping out of sight.

They lacked a bear's full share of caution, though, and soon reappeared in Walt Schock's field. Hungrier than ever, they resumed the work of scenting burrows, tearing up sod, and bolting down gophers.

The cubs stayed put after that, and Stacy did all that he could to trap them. He had to be careful in his efforts. Many bears were out and active — all looking for a meal — and Stacy hoped to avoid the trouble and danger of catching an adult grizzly. To this end, he chose to use a smaller trap made of welded steel mesh, one built to catch mountain lions. A smaller adult bear could still fit in there but not easily, and Stacy took a further precaution by trapping only during the day. At night, when mature bears were active, he tripped the trap's gate, locking it shut.

It went on for a week. The cubs were often out feeding. Stacy watched them, finding that they would allow him to approach closely in his pickup truck. Sometimes he narrowed the distance to ten paces. Then he was near enough to see their pupils and the small movements of their ears. He would have noticed that the cubs were particularly beautiful bears, each having a blond mark that looked like a kerchief running down from their hump.

He saw, too, that one of the cubs was larger than the other and that the more diminutive cub was unwell. The larger sibling dug and hunted constantly, relentlessly, but the smaller one would sometimes give up the work, walk off a distance, and sit motionless. There was a lethargy in the little bear that worried Stacy, cementing his conviction that the cubs had no hope of surviving in the wild.

The cubs let him draw tantalizingly near, almost close enough that he could reach them with a long-handled catch pole. Stacy considered trying to dart them from the truck. The thought tempted him, but he put it from his mind; it struck him as too risky. A free-darted bear — one hit when it isn't caught in a trap or otherwise immobilized — can run for several minutes before the tranquilizing agents take effect. Those minutes are enough to take a scared cub far away and deep into an impenetrable thicket. Concerned that he might not be able to dart both cubs at once or recover them once anesthetized, Stacy pinned his hopes on the trap and waited.

He used all sorts of odiferous baits, but the cubs wanted none of it. Labor Day weekend arrived, and Stacy kept on with his work. Heading down to check the trap on his holiday, he was surprised to find the cubs absent from the field. Sudden absence augurs poorly for orphaned bears, and Stacy must have considered the prospect that a boar had found and dispatched them.

Idling along in the pickup, though, he found that the cubs hadn't gone far. Just down the road, he caught sight of them in an orchard adjacent to Post Creek. The trees were laden, and the cubs were gorging on fallen fruit. They retreated when Stacy towed the trap down from Walt's field and set it up in the orchard, but they were not long in returning. Stacy baited the cage with apples and had caught one of the cubs within an hour.

A friend of mine who works as a game warden tells a story about a run-in with a little bear. Early in my friend's career, a biologist had asked him to remove a black bear cub from a holding pen. The beast was petite and cute. "Raccoon-size or smaller," as the warden put it.

He went into the cage feeling sympathetic, thinking *teddy bear*. He came out empty-handed, lacerated, and humbled. When he had made a grab for the cub, it stuck to him as if they both had been made of Velcro. He said that the cub moved faster than he could follow, clawing and biting, and that picking it up was like sticking a hand into a garbage disposal.

Knowing that the larger of Millie's cubs weighed nearly fifty pounds in September of 2016, I can imagine that the work of transferring the caught cub to a culvert trap—a task for which Stacy enlisted Shannon's help—required no small amount of skill and concentration.

Rebaiting, they caught the second cub just as darkness spread across the valley. Bumping away from the pasture to the sound of claws scrabbling against metal, Stacy and Shannon hauled the terrified cubs to the office of a local veterinarian.

They worked them at the vet's place, tranquilizing, weighing, and measuring each animal. Both bears were female. The veterinarian x-rayed the sisters, and an image of the smaller one's hindquarter revealed a constellation of bird shot embedded deeply in her muscles. Without any sign or report of a sow, Stacy decided to haul the cubs to a Montana Fish, Wildlife & Parks rehabilitation facility. He made the drive without delay, speeding the cubs through the Mission Valley for the last time in their lives, depositing them in an electrified pen on the outskirts of Helena to await their fate.

They could not stay there for long. The Helena complex is the state's sole place capable of holding bears, and it has only two pens.

The cubs had one of these to themselves, but things were about to get crowded. Fall was coming on, and hunting season always creates a rash of bear conflicts. More animals would be brought to the facility, and the cubs could not share space with them. The sisters' prospects were bleak if their mother could not be found. Either a zoo or sanctuary would agree to house them permanently, or they would be killed.

I SHOWED Stacy the footage that I had captured of a wounded bear in the cornfield. Upon watching the video and studying a pair of freeze-frames that showed the damage to the sow's face, he said gravely that he had never seen a bear look quite that way before. Without delay, he shared the images with other biologists at the tribes' Natural Resources Department and Montana Fish, Wildlife & Parks.

Experts were soon weighing in with theories about what had happened to the bear. A veterinarian employed by the state watched the footage and said that the bear might have cancer or a strain of mange. Stacy said that it looked to him like she'd been shot in the face and that it wouldn't have been the first time a bear had met such an end near Millie's Woods. Because countless pheasants and partridge lived in the Mission Valley's wheat and cornfields, bird hunters plied the area around Post Creek. Dancing on the razor's edge of foolishness, they toted shotguns along overgrown fence-rows and through boggy swales.

Game birds and grizzly bears enjoy some of the same habitats, and both species made use of the agricultural fields around Millie's Woods. This overlap had resulted in a number of nasty encounters between bears and sportsmen, as well as countless close calls.

Because of this history, the tribes had taken to closing all the land they owned in and around Millie's Woods from September 1 through November 30 each year. The closure does a great job of keeping people from wandering into the heart of the woods when bears are most active. It does not, however, apply to the surrounding private land.

"There's nothing we can do about them hunting on private ground," Stacy said, and he went on to describe the lack of concern that certain hunters displayed when it came to grizzly bears.

The year before, while making rounds with his radio receiver, he had tracked a collared sow to a cornfield in the valley.

"Not Schock's," he told me. "But it was the same kind of deal. I was in my pickup waiting to see what the bear would do when a truckload of hunters pulled up."

Getting out of their vehicle, the men began to load shotguns and don orange vests, making preparations to enter the stand.

"There's a grizzly in there," Stacy told them, but the hunters were unimpressed. Despite his warning, they went on sorting out their gear, then disappeared into the corn.

They hunted that whole field. Telemetry unit in hand, Stacy listened as the bear moved back and forth, always keeping ahead and away from the hunters. He was impressed by the sow's ability to steer clear and frustrated by the hunters' lack of common šense.

"Those guys got lucky," he said. "Not everybody does."

He explained how easy it was for a grizzly and a hunter to surprise each other — an encounter that often led to a frightened gunshot.

"Is that what happened to the bear in the video?" I asked him.

"Could be," he said. "The wounds look right."

STACY HAD something in common with the animals he managed. Like a bear, he spent most of his time moving cautiously, even ponderously, through the world. He chose his path with care, seeming never to hurry. But he could charge when the situation called for it, and as September came into its crisp fullness, he focused his mind and efforts on the wounded bear and Schock's field, pressing his cause with striking determination and speed.

I focused on Millie, too, though I did not know her name then. I cared for her the moment I saw her ruined face. She was damaged, far gone, and ghastly, but I was instantly attached to the sow.

Part of it was guilt. She was inside my fence and had looked out across the wires in such a way that I could not help thinking she wanted to cross them. Another part of it was the horrendous condition of her face, which embodied the violence that humankind inflicts everywhere on wilderness and wild creatures.

I was nothing to the sow, but she was something to me. From the moment I saw her, she struck me as exceptional. Without meaning to, I came to have an interest in her fate.

Though I had no hard evidence that the ruined bear was female, I theorized that the cubs Stacy had trapped were her offspring. Deciding that my fence had separated a mother from her daughters, I hoped that their story would end with rehabilitation and reunion.

I DROVE up to Schock's field from Missoula on a Wednesday morning, passing through Wye before the rush and climbing Evaro Hill against the flow of commuters making for town. Coming to the field with the first sunlight, I found Stacy and Shannon waiting on Hillside Road. Standing beside their three-quarter-ton Ford, I leaned on the cab and looked at the brown-leafed corn.

"I worry that she's stuck in there," I said, "and can't go where she needs to."

"She won't leave," Stacy said when I proposed shutting off the current and trying to run the wounded sow out of the corn. "I'd bet on that, long as she has food and water. She won't come out, and if you turn that fence off for even one night, the field will be as full of bears as it was last year."

The only option was to trap her, and the men had been trying to do so for a day or two already. They had left one of their larger culvert traps in the pasture overnight and were keen to see if bears had visited it.

"If you catch her," I asked, "will you put her back with the cubs?"

The question took Stacy by surprise. "Put her back? No. The way she looked, I don't think we could do anything for her."

He leaned forward and glanced through the windshield at the sun, which had climbed free of the peaks into the bowl of the sky.

"Probably bedded for the day," he said to Shannon, who nodded in reply.

Stacy turned to me. "Want to come and check the trap?"

There was no room in the cab of their work truck — every inch was crammed with maps, tarps, dart guns, shotguns, and other paraphernalia necessary for securing and moving dangerous animals — so I offered to ride in the back.

The pickup rattled along the road. At a gate, I jumped down, shut off the current, and pulled the wires aside. When the truck had crossed through and stopped, I set the fence right again. Stacy drove slowly along the edge of the corn. Sitting on the rim of the bed, I watched stalks passing close enough to touch.

Stacy followed the only navigable path into the field. Beaten in and kept open by the tires of Greg's rig, that two-track was so narrow that I had to dodge and bend to keep clear of overhanging leaves. Looking across the tailgate, I could see parallel walls of tall brown plants and a lit rectangle opening to the outer world. It shrank, and as we passed into one of the field's hollows, it slipped entirely from view.

Catch poles, tins, and toolboxes framed the bed's main pay-load: a bloated, reeking whitetail carcass, its hind legs shattered by someone's bumper. The deer shook and slid as we bumped across ruts. Edging away from it, I knew that the smell of death was moving through the field as we drove on, spreading like chum.

Standing with my feet braced wide, I looked across the cab's roof. Ahead, I could see the pivot's center and a thick vertical main line rising fifteen feet to meet the sprinkler's first segment. Nozzles hung dry above the tassels of the corn, and beyond them, the stand gave way to alfalfa. Out there, just beyond the final stalks, sat the trap.

The two-axle trailer was black, with a drop-down ramp of perforated metal at its back end. Its deck was four feet wide between the wheel wells and perhaps fourteen feet long from hitch to tail-lights. The style and shape of the flatbed were not new to me, as I had used similar trailers to haul four-wheelers and snowmobiles in the course of ranch work.

My sense of familiarity stopped at the heavy worn boards of the decking. Atop them sat the trap, a long octagonal tube of brushed aluminum. Small holes, each an inch in diameter and showing black in the morning sun, had been drilled through the metal at regular intervals. The trap was plastered with signs reading DANGER and appended at either end with vertical barred gates. The front gate was shut, but the back one had been hauled upward into its housing by means of a cable running to the contraption's center. It had an ominous look, like a guillotine primed to fall.

We idled toward the trap's open end, and Stacy peered within. Satisfied that nothing lurked inside, he cut the engine.

"This," he said when I had climbed down from the bed of the truck, "is a culvert trap. We call it a family trap, because you can rig it to catch a sow and cubs at the same time."

Walking to the big cylinder, he pointed out a third gate midway along its length.

"Shut this," he said. "And you can get a bear in either side. Hardly ever works out that way, though. Mostly you catch the sow, but the cubs won't go in, or you get one cub and not the other."

"Or," Shannon broke in, "you get the cubs and the sow stays loose."

"Right," Stacy said. "Last year Shannon was checking the trap, and about the time he realized he had both cubs inside, the sow came blowing out of the brush — she was close, like the corn is there — and damn near had him before he could get to the cab."

"Could've had me if she'd wanted," Shannon said, grinning.

Walking a short way off from the trap, Stacy studied the ground.

"She's been around, or one of the other ones has," he said, indicating a pile of corn-packed scat with the toe of his boot. "This wasn't here yesterday."

Coming back over, he bent to study the ramp's honeycombed metal, which had been scattered thinly with hay.

"She never put a foot on it," he said.

From where I stood, the trap looked as unnatural and ominous as anything could. Watching Stacy fuss beside the gate, I struggled to imagine a bear venturing inside.

"Do you catch many this way?"

Stacy quit working.

"You've got to understand how the world looks to a bear. You and I wouldn't crawl into that thing, but a bear likes a tight spot. He spends his winter in a hole and goes to the thick stuff if he wants to rest. From his point of view, a trap like this doesn't seem bad as long as it smells right."

I thought about the ursine nose and how bears had been observed walking miles upwind in a straight line toward a carcass.

Feeling sure that any boar or sow in the field could scent my breath coasting south on the breeze, I took a step away from the corn.

"It's got to smell right," Stacy went on. "Because a bear's nose runs him. We know that, so we're particular about what we put around the trap. See, Shannon's setting out sweet feed."

Turning, I watched Shannon sprinkle handfuls of molasses-soaked oats across the ground. He worked carefully, laying a rough trail of grain between the standing corn and trap. Walking to him and looking closely among the alfalfa plants, I saw that other attractants — scraps of bloody hide and small bits of flesh — had been set out as well.

Going to the truck, Shannon returned with a squat plastic tub in hand.

"What it comes down to," he said, unscrewing the lid, "is that we use whatever works. They make all sorts of lures for trappers — beaver lures and bobcat lures. We're always looking to find out which ones the bears like, and this stuff, this 'Horse Whisperer'" — he tapped a finger against the plastic canister in his hand — "works.

"Smell it," he said. When I hesitated, he held the tub to his own nose and sniffed.

"Really," he stuck out his hand. "Go on and smell it. It's not bad, like some of the others."

I dipped my head, took a whiff, and found that he was right. The thick reddish paste smelled rich and funky, like rotting hay or leaves on a forest floor.

"Horse Whisperer?" I said. "Meaning it's made from horse meat?"

"Meaning horse meat's in it, and a lot of other things. But it's the good stuff, whatever it is."

Stooping, he felt among the trampled grass. Finding a rigid stem, he fished it around in the tub and extracted a clot. The glop amounted to no more than a teaspoonful, and Shannon handled it carefully, spreading it over the ramp's expanded metal as if he were buttering bread.

"They stopped making it last year," he said wistfully. "We laid in a supply."

Digging loose another wad, he smeared it onto the metal just inside the mouth of the trap. Peering past his hand, I saw the far gate ratchet up and Stacy's wide shoulders fill the opening. He spoke through the trap's long tunnel.

"Course you need something for the bear to eat once it gets in there. You need something that'll bring it to where it can trip the trigger. And that's" — he lifted something large and dark, and set it on the trap's rounded floor — "why we pick up roadkill."

"Green-bellies," Shannon said. "Ripe ones. Sometimes you can't hardly breathe by the back of the truck."

Stacy disappeared from view, the stink of death reached us, and after a moment, the trap's far gate clanged down. Shannon rearranged the straw covering the ramp and headed for the pickup. Following, I climbed aboard and settled in to ride. We bounced out of the field on Greg's access road with leaves hushing against the side panels. Glancing at the whitetail carcass — now missing a hindquarter and oozing blood onto the Ford's white corrugated metal — I wondered what the sow would do.

WITH THE exception of certain cold mornings, summer held on. The sun gathered the valley in its baking embrace, moisture vanished from the soil, and creeks dropped as though a hand were shutting valves far up in the mountains. There was a place where I liked to swim on my way home from working at Schock's field, an eddy where the Jocko River ran black and slow. All through August, it had remained deep enough to dive into, but in mid-September, the pool dropped until I could see bottom. The water grew colder, too, so I could do nothing but dunk, scramble out, and shiver.

Hauling myself onto the bank, I lay down in a sunny spot. The wounded sow was on my mind, which was nothing new. Ever since visiting the trap, I had been plagued by her. I chided myself, asking what right I — hunter of deer and elk, hauler of cattle to slaughterhouses — had to be sentimental about the life of a single grizzly.

I couldn't help it. As I rested on the bank of the Jocko, eyes shut against the afternoon glare, the sow was with me. I lay still with the sun glowing through my lids, and she hovered above my upturned face.

I tried to put her from my mind, but the sow remained in my incarnadine darkness, showing her missing tissue and sores. Worst of all was the fact that I could do nothing for her. Stacy and Shannon couldn't catch her, though they tried all sorts of baits and scent lures. Even if they did manage to trap the sow, I knew that I would have no say in her fate.

The sun lowered by the minute, shadow fell over my legs, and I tried to think of something else. I made it as far as the orphaned cubs, wondering what sort of life they were leading in Helena.

Then the grisly vision was back, the sow looking in my direction, displaying her wounds.

Seeing her that way built up a reservoir of feeling in me — a dammed lake of care.

TAKING A personal interest in the bear's plight, I went daily to Schock's field. If I rose early enough, I found that I could cross paths with Stacy and Shannon when they came to refresh the trap's bait. On the fourth morning, I met the men at the field's southeast corner. They were leaving empty-handed, and coming abreast of one another on the gravel road, we shut off our engines and talked.

"She'll come right in close to the trap but won't touch it. Too much food in the field, or she doesn't like the look. But we'll catch up with her soon." Stacy nodded westward. There was nothing that way but hazy sky and acres of browned-out corn. "Greg will be chopping soon. A couple more days, one more frost, and he'll start cutting. Then those bears will have to move, and one way or another, we'll get a look at her."

Putting his truck in gear, Stacy drove north toward Post Creek. Parking, I started walking clockwise around the fence. I meant to cover the whole perimeter to make sure that the current remained strong and check the condition of the wires. Knowing that the wounded sow and at least one other bear were in the corn, I kept outside of the wires.

I headed west from the southeast corner. A wide shallow ditch met the fence on that side of the field, and over the course of the summer, I had found just one good place to cross it, where hummocks of grass stood above the flow like small islands. Coming near the ditch, I detoured out of habit. When I reached the crossing, though, I found it dry, with the bottom mud already curing.

Following the fence, I cleared away bits of new growth and windblown skeletonweed from the bottom wire. Before reaching the southwest corner, I knelt to test the current. Taking a reading of 7.1 kilovolts from the top wire, I stood up satisfied.

Beyond the corner, I came to where the dried-up ditch left the field. There, its channel grew narrow and was gouged two feet deep in clay-rich soil. The spot felt like an oasis, and the turf for several yards on either side was spongy.

The tracks were what I noticed first, wide ovals where my fence crossed the waterless ditch. Through August, when the little canal had been flowing, the lowest wire had hung just a few inches over the waterline. Now it was two feet above mud, and a bear had taken notice.

Coming closer, I could see that the floor and sides of the ditch were crosshatched with claw marks. Stooping, I could see clearly how the thing had been done. A grizzly had pulled soft earth and cobble back along the channel, excavating a trench deep and wide enough to squeeze through. Pressing into the mud, fur leaving a mark like the stroke of a huge paintbrush, it had wriggled beneath the fence without touching wire.

Crawling through the hole myself, I picked up the tracks where they climbed out of the ditch. Heading for the stand, the bear had passed a low place where the field's runoff had pooled and remained. That green stagnant water hole was ringed with older prints, all preserved as faithfully as if they had been cast in plaster. The tracks came from all directions, and I supposed that what grizzlies were in the corn — three that I knew of, including a newcomer with a talent for digging — came to the pool to drink.

Examining the tracks, measuring them against my hand and one another, I tried to decide how many different animals were represented. In the end, I could determine only that bears of several sizes were in the field and that one of them left tracks longer than a size-twelve boot.

Resolving to return and set up a trail camera at the watering hole, I crossed back to the ditch. A heap of glacial stones sat near the place where the bear had tunneled under. Some of those boulders, which had been exhumed over years of farming and skidded from the field, were too big for a man to move. Others were small enough to tempt me, and lifting one, I staggered toward the spot where the bear had entered.

I piled many stones at the ditch's edge. The sun burned defiantly against the changing season, and I sweated through my shirt. Judging that I had enough to work with, I climbed into the channel. Kneeling, I used my fingers to excavate hollows in the banks and bottom, obliterating the claw marks that had been there. Laying a foundation of larger stones, I built a wide low wall across the ditch. I took my time with the task, fitting each rock closely beside its neighbors and orienting the flatter sides to present a tightly jointed face to the outside world.

Caked in mud and sore-handed, I examined the stacked stones from several angles. Though I knew that the wall could not stand long against a determined bear, I judged it to be enough of an obstacle to give trouble. A grizzly would have to labor at getting through and would probably be shocked by the fence's bottom wire.

One particular rock in the wall caught my eye. I had found it in the bottom of the ditch, where it had been dug up and pushed aside when the bear came through. It was roundish, slightly smaller in diameter than a basketball, and slick with muck. The grizzly had moved it once, shoving it with a paw, plowing a furrow. I had moved it back with muddy hands.

We both knew the thing's weight and balance, and there was something intimate in that. My fence had made the bear dig. The creature, in turn, had forced me to make repairs. Laboring in the same spot, in opposition to each other, we came away painted with matching soil.

DEEP IN the rows, injured and wary, Millie kept hidden. She moved mostly at night. Some moisture could be got from the corn, but she needed more to survive and had to visit the shallow pool of standing water near the field's west fence. By the middle of September's third week, that oasis had dried to a handful of muck-ringed, scum-flecked puddles. It smelled of unwholesome rot and agricultural chemicals.

Millie drank there each night, and the wounds around her nose, eyes, and mouth worsened, making it harder to chew and see. In spite of her condition, she made a short daily circuit — from corn to water and back again — as best she could.

Grizzlies are marvelous in their resilience. I heard once about a bear that had most of its lower jaw shot off by a hunter, so only its back teeth and mandible remained, and its tongue hung out naked to the world. It was horrible to see, and everyone with a right to an opinion thought that the bear couldn't last more than a few days. But the creature, eluding all the traps that were set for it, stayed alive. It survived for weeks, feeding by smashing apples to pulp with a paw, then slurping up the bits with its tongue.

Millie was no less damaged than that bear and just as determined to keep living. She ate what she could while rot consumed the tissues of her face. Infection caught and flared, ravaging her and undermining all her efforts.

She was never free from pain. For a creature with a shattered nose, the cornfield was a terrible place to be. Walking anywhere meant pushing headfirst through inch-thick stalks and sharp-edged leaves.

The corn, which had fed her well for years, did not keep her from dwindling. She ate all that she could, but because of her injury and the infection sapping her energy, the crop could not sustain her.

Day by day, she wasted in a field of plenty. All the while, she refused to enter the culvert trap, though Stacy and Shannon tried all manner of baits.

Perhaps this was because she could not see or smell their various lures and offerings. Stacy once caught sight of her near the trap. Watching from the cab of his truck, he saw her walk slowly from the water hole toward the corn. Oblivious to the presence of both trap and truck, she vanished into the stand without a glance or sniff in Stacy's direction.

I CAME five days in a row to the corn, checked the cameras, and saw nothing but a great many tracks. The bears moved like ghosts. Even when I flew the drone overhead, I could not see them. And so, when my parents drove out from Seattle for a visit, I took a break from work.

Though they raised a feral son who departed for the hinterland at his first opportunity, my parents are urbane. My father, Richard, ran a contemporary-art museum for most of his career. Though he is now retired, he once shepherded divas and curators better than I have ever wrangled calves and horses. My mother is a photographer by trade, an artist who has made her living documenting art and architecture in the foremost cities of the world.

All this is to say that the particulars of my life and work are more foreign to them than the streets of Milan or Buenos Aires. They love me fiercely, though, and no doubt worry more than they let on. We passed a fine weekend together, and Monday morning brought a hard frost to Missoula. We went out walking in the clear sun, the three of us moving with the same long strides on trails through the hills near town. As we hiked, they peppered me with questions about the cornfield.

My mother asked how grizzlies ate the corn, and I showed her, hunching ursine beside the path, opening an ear with long claws, and gnawing. My father asked what I thought about the farmer growing corn so close to the mountains. I said that it was complicated.

He asked if the corn was organic. I told him that it wasn't and that Greg sprayed the field regularly with herbicide. When my father wondered aloud what that did to the bears, I didn't know what to say. The question is a tough one because bears eating corn is a new phenomenon.

But glyphosate is a potent substance, and the bears that gorge, drink, and bed in a sprayed field are exposed to a lot of it. Because the chemical works by disrupting an essential metabolic process in plants—an enzyme pathway that does not exist in animals—most chemists consider the compound safe for humans, cattle, grizzly bears, and other mammals. "Less toxic than table salt," I've heard herbicide dealers say.

I doubt this claim and am not alone in doing so. Glyphosate is used everywhere in modern agriculture and now shows up in human urine and breast milk. A doctor friend of mine maintains that its effect on our intestinal flora—some species of which metabolically resemble plants—is poorly understood and worrisome. Having read and written extensively on the subject, she now eats only organically produced food.

Others are concerned, too. A 2015 study by the World Health Organization's International Agency for Research on Cancer classified glyphosate as a carcinogen. Though the finding was hotly disputed by other regulatory agencies, the United States Environmental Protection Agency among them, I still want the stuff kept out of my food.

The impact of glyphosate on Mission Valley grizzlies remains unknown. There can be no doubt, though, that the nutritional profile of corn reshapes the minds and bodies of bears. Packed with starch and sugar, the grain allows grizzlies to put on massive amounts of weight without expending much effort.

Bears grow larger and less mobile when they have access to corn, becoming what Stacy calls "belly draggers." The previous year, he told me, one boar quit the crop only as Greg's tractor cut down the last row. The animal had just half a mile to run, but it stopped twice to pant before crossing out of the field. Its efforts were pathetic, and Stacy chuckled when he recalled them. I had never before heard him laugh at a bear.

Too much corn ruins a grizzly in the same way it ruins human beings. The bears get too fat — far exceeding the weight gain necessary to survive winter — and their teeth fail. Bears frequenting the field are more prone to dental plaque and decay. A seven-year-old corn-fed bear, for example, might live with the abscessed molars of a fifteen-year-old.

The crop replaces a dozen varied seasonal resources with junk food. It strips away self-reliance and the need to move, turning a mountaineer into a sedentary creature. In doing this, corn leaches the wildness out of bears, making them too much like us.

STANDING IN dry grass at a place where we could see Missoula covering the valley like a feast spread across a table, I talked with my parents about the difficulties of summer and how the bears, desperate for ripe ears, had learned to jump the fence at pivot crossings. I told them what it was like to work beside the field when grizzlies were in it, how I had come to know what a deer feels when its head and tail come up simultaneously and everything in it tenses.

"It sticks with me," I said, feeling the simmering primal alarm still present as a low-grade ache between my shoulder blades.

I tried to explain that it was not fear I felt beside the corn but a mortal, ancient form of concentration. Beside the sword-leafed wall, my mind and senses were like sunshine passed through a magnifying glass.

It felt good, or at least natural, to be like that. At the field, I didn't think about the future, or think at all, really. I just worked and watched, paying attention to tracks and the wind, to noises and smells.

Mom stood looking south, face hidden by her hair. Adjusting the strap around her neck, she shifted her brick of a camera from one hip to the other.

She said: "That wounded bear, the one they were trying to catch?"

"They're still trying."

"What about her cubs?"

"Nobody knows for sure that they're hers," I said. "But they went to the Fish, Wildlife & Parks facility in Helena."

She asked what would come next for them, and I said that they would go to a zoo if one could be found or be killed. The news did not sit well with her.

Blowing from the east, the wind carried a new chill. My dad stirred against it, shifting his weight from leg to leg.

"We should go up," he said. "If you've got time tomorrow, we should go up to the field and see."

Walking downhill, we talked of other things. At home, I found a voice mail waiting from Greg Schock.

"Hey-there-Bryce-Greg-here," he said. "Had a frost last night, and I haven't seen you these last few days. Well—I know you wanted me to get in touch when we started on the corn, so I'm calling to say we're chopping this afternoon, and we'll be at it the next few days." He paused long enough to take the phone away from his mouth and give a muffled order. "Go on and hitch it up," I thought he said, then his voice returned more clearly: "I'm going to shut that fence off while we're working, but I'll keep it on at night. You wanted to see bears, and now's the time, or pretty soon it will be, once we get most of that corn down."

A tractor's engine rattled to life in the background. Without a pause, Greg said: "All right, c'mon and back it up, back—all right, *all right,* hold there—" And the phone clicked off.

A NERVOUS enthusiasm crept into both of my parents overnight. Standing in the driveway, my mother arranged her camera gear, changing one lens for another and checking batteries while I loaded up the Subaru. Dad stood ready by the door, checking the weather on his cell phone. They were eager, excited at the prospect of seeing my work and the grizzlies.

It was midmorning by the time we headed north, and we crossed into the Mission Valley with questions bouncing around the car. Suspecting that we were headed for a letdown, I said: "More than likely, we won't see anything but corn and Greg's hired guys chopping it."

Coming into the field, I could see the harvesting machines from a long way off. They were close to the field's edge—the tractor leading with its loader forks high and the chopper dancing along at the end of the PTO. The harvester was worn, with a wide green duck bill that swallowed three rows at a time. Behind it trailed the silage wagon, a high four-wheeled hopper collecting a steady stream of minced leaves, stalks, and ears.

The wagon's tires belled out as the corn piled up, the complaints of its suspension mixing with the harvester's furious sound. The machinery rode a rising tide of noise, and when the sickle met a clod, dirt shot up as if the convoy had struck a land mine. Rising high, dust obscured the red shoulders of the hopper before dispersing in an unclouded sky.

The tractor looked as small as a toy, and I saw Greg's old dump truck heave into view and pull alongside. The two machines traveled in parallel for a moment before the tractor driver stopped. Up came the hopper, rising on hydraulic cylinders until silage cascaded into the dump truck's bed. Then the hopper settled back, the tractor belched a black plume, and both drivers eased their machines on.

Coming to the end of the row, the tractor bent into a wide right-hand turn. Crossing the field's dirt rind, it began another pass. Parting ways with the chopper, the truck rolled heavily through the gate and headed for Greg's dairy barn.

When the truck had gone, I could see that the tractor driver had already opened a wide gap in the corn. The cut space covered a third of the field, and south of it, a long strip of standing corn had been separated from the main body of the crop.

A reversal had taken place in my absence. The field, which had seemed as endless as an ocean, was now split into a pair of islands. The tractor circled the smaller stand, each pass diminishing it by three rows and widening its separation from the rest of the corn.

Though the machine moved no faster than a walking man and cut just six feet at a time, the driver had put in many hours. His work had the inexorable quality of erosion. At a hundred yards across, the harvested space was twice the width of the uncut strip.

Stacy's white truck was parked at the far end of the newly opened space. Pulling through the gate and taking a bearing on him, I steered the Subaru across crew-cut stubble. The stalks, which looked inconsequential from the road, proved to be eight inches high and rigid enough to tear at the underbelly of a low vehicle. We drove through the field with a terrible noise, and though my father wore the tight look of a man witnessing wanton mechanical destruction, he said nothing as we crossed the field and bumped to a stop near the Ford.

"Morning," Stacy said when I got out and walked to his window. Shannon nodded from the far side of the cab. "Rough driving. How's things?"

I told the men that I had brought my parents up to show them the field and perhaps a bear.

"Might see plenty today," Stacy said, and looking into the pickup, I could see a scoped rifle resting against the center console. I waited for him to go on, but it was Shannon who broke the silence.

"There were two in the patch he's cutting now. Two at least. One made a run for it, and the other stuck its head out a minute ago."

"Last night," Stacy said, "the driver ran across that sow. She was in rough shape is what he said. Could hardly keep ahead of the chopper."

Without meaning to, I looked at the rifle. Stacy answered the movement with a nod.

"If she can't stay out of the way — you see how slow the chopper goes — she's done."

"So," I said, "you'll wait until she crosses?"

"If she's in there. It was last night he saw her, and a lot can happen in a night. She could have crossed to that bigger patch. She could have died. But if she is in there, and if she goes, and if I get a shot — Shan, you see that?"

"Sure," Shannon replied without taking his eyes off the outermost row of corn. "Stuck his head out again."

I strained my eyes, seeing nothing but dry leaves and stems. After a while, Stacy turned back my way.

"You bring that drone today?"

I told him that I had.

"Couldn't hurt," he said, "to get it up and see what you can see — if there's more than one bear left in that patch he's cutting."

SEEN FROM high above, the crop was profoundly changed by the late season and frost. Leaves had blanched to the hue of parchment and wilted until dirt showed between stems.

I flew slowly up the strip of corn with the camera pointed straight down, making an effort at a systematic search. At the field's far end, I hovered while the harvester made its turn, then eased the drone forward with nudges of the joysticks. I stayed just ahead of the tractor as it drove west, keeping the drone fifty feet off the ground.

The corn had been whittled thin. The stand's whole width was visible in the camera's frame, and I doubted that any bear would stay in it with the harvester's racket less than thirty yards away. I squinted against the glare while my father, watching across my shoulder, shaded the screen with his hands. Turning once, I saw my mother near the car, photographing bear prints in dry mud.

Back at my work, I edged the drone forward every time the tractor entered the lower corner of the frame. For what seemed a long while, I saw only gridded rows of light-brown leaves. Then came a flash of motion. Stalks stirred and parted like water breaking over the back of a spooked fish. The commotion shot forward, and I accelerated to keep pace.

The disturbance vanished as quickly as it had appeared, and I was left hovering over what seemed to be an empty patch of corn. I would never have seen the bear again if it had not stepped into a void where stems had been trampled, paused, and raised its head to study the craft hanging in the sky.

The head came up dark and unmistakable among the tassels. Round and thickly furred, it tracked the drone like a flower following the sun. Descending, I could see the bear in detail. He was enormous, and his lips were white with kernels and saliva. Bobbing his head slightly, he stared upward with small obsidian eyes.

Dropping his head, the bear vanished. No shake or ripple in the stalks betrayed his direction, though I watched until the tractor labored up and trimmed another three rows from the corn.

I flew until the machine was out of power, changed to the second battery, and wore it out, too. I saw the bear twice more.

The first time, he appeared in a swale where standing water had drowned much of the crop, leaving the cover patchy. The second time, I glimpsed him standing so close to the chopper that I could hardly believe it. He stood his ground until the cacophonous thing was upon him, until he looked like he would be caught in the cutting teeth, before sprinting away.

Bears would let the machine pass very close, but they did not like it. As the chopper pared away the stand's edges like a hand plane smoothing wood, it also worked on a grizzly's mind. Every round brought the equipment closer. The smell of diesel and the roaring of engines were on the wind. Metal fingers jostled the stalks with implacable strength. The harvester crept on, and bears moved fitfully, losing the sense of security they had known in the corn, crossing and recrossing the stand.

Having exhausted the aircraft, I led my parents to where Stacy and Shannon waited. We talked until the sun was high and the tractor driver had whittled the corn to almost nothing. Though still as long as the field, the stand could not have been more than twenty yards wide.

"He's got to run," I said, and Stacy nodded in agreement.

Shannon toyed with a pair of binoculars on the dash. He cleared his throat.

"Funny how close they'll let that tractor get."

The two men launched into a debate about whether the injured sow remained in the narrow strip, settling finally on the theory that we would have seen her by now and that she had probably retreated to the larger remaining patch.

Still, tension built among us as the stand thinned. My mother, never straying far from the truck, photographed cut stubble and the boggy place where bears came at night to drink. Dad, Stacy, and Shannon kept a steady watch on the outermost stalks.

My father, after watching the empty field for a very long time, stepped behind the truck to pee. He had not been gone twenty seconds when Shannon said, "There! There's his head again."

A dark spot showed among frost-killed leaves perhaps a hundred yards east of where we stood. Then the bear was out and moving, stark against tan stubble, sprinting across the cut field. As he gathered himself and stretched, taking fifteen feet at a stride, I heard the shutter of my mother's camera. In seconds, the bear was gone and the corn stood as motionless as ever.

"My god," Mom said.

Stacy looked at her. "They're fast when they want to be."

"Good-size bear," said Shannon, and we turned to see my father returning along the side of the truck.

"Shit," Dad said when we told him, and he settled into a fierce vigil, as if the bear might dash back in the other direction. He kept it up until the small patch of corn was diminished to six rows, daylight shining through, and the biologists took a new position on the field's far side. Only after they had pulled away and I had started with Mom toward the car did he reluctantly turn his eyes away from the field.

We ate a late lunch at a diner on Highway 93, and while we waited for the food to come, my mother said: "I didn't know they could run like that."

"Like what?" Dad asked.

"So fast. And when they run, they change. The fat goes away, all the slowness that you think of when you think about a bear. They're all muscle and strength, like an enormous dog."

COMING BACK to the field, I made for its north end to show my parents the pivot crossings and change the memory cards in the trail cameras. Seeing no sign of Stacy and Shannon, I figured they were on the far side of the remaining corn, waiting and watching as the tractor made its rounds.

The chopping crew had been busy while we ate. The stand's southern edge had receded, and the driver had already made a single pass beside the narrow road that Greg used to access the center of the pivot. His work had divided the stand again, though the swath of cut stubble was only twelve feet wide. Staring as we drove past, I thought it strange to see a sunny, trimmed path running toward the heart of the corn.

Parking at the field's northwest corner, I stepped from the Subaru into a day that had grown warm. The sun was in the western sky, and my father shaded his eyes against it. I clipped bear spray to my belt and dropped three fresh memory cards into the breast pocket of my shirt.

We went down the line together, walking in high grass just outside the fence; and looking over my shoulder, I could see doubt and fear in my mother's eyes. She watched the stand as we neared the first camera. Stepping across the fence to reach the device, I could hear the chopper working far off, its engine noise punctuated by the clash of steel on steel.

We went farther down the line, my parents hanging back a few steps and talking in low voices. Ahead, the second camera stood on its post beside the corn. The chopper sounded closer, and I guessed that the driver had made his turn at the field's western edge. He was cutting along Greg's two-track access road, no more than fifty yards away from me in the stand. Realizing that the harvester could spook a bear in my direction, I paused midstride.

"Is this safe?" My mother's voice was small beside the vastness of the corn, and I thought for a moment about how to answer.

"Not really," I said.

We reached the third camera. Pointing out a bear track in the pivot's rutted track, I stepped across the wires into the cornfield.

The camera was set well inside the fence; and walking to it, I maintained a steady conversation with my parents. With the corn so high and close, I was glad to have their company.

"Don't you worry," Mom said, "that a bear will come out?"

"I do, but now that I've seen the bears with the drone, how close they'll let the tractor get, and how much it takes to make them run, I think they just want to stay in the corn."

"But you're not in a tractor. You're just standing there."

I switched out the memory card and shut the camera's plastic clamshell. My hand was still on it when a gunshot rang out close and loud. A crisp, hollow noise, it slapped hard against my chest. *Rifle*, I thought, trying to peer through the nearer rows. A tumult began within the corn, and as I backed away, redwing blackbirds took flight by the hundreds. Rising with a thrumming, cresting noise, they banked east as one and fell like stones.

The second shot pealed out as I was hurriedly crossing the wires. It was as loud as the first; and on the heels of the report, I could hear the dead-dull *thwap* of a bullet striking home.

"He hit her," I said as I reached my parents, and we beat a hasty retreat up the fence. The harvester's noise ceased. Stillness descended. We had nearly reached the car when a third shot broke the calm. It was fainter and flatter than the other two, and I thought that it sounded like a pistol.

We sped along the road. Turning in through the field's main gate, we bounced to where Greg's road entered the stand. When the dump truck heaved into view, I hailed the driver.

"They got her?" I asked.

"Sure did," he said. "Right off the road."

I steered the car down the ruts, though I knew that the access road was no place for a station wagon. The swales were still muddy, and I avoided them by perching one tire on the high center crown of the road and bouncing the other over cut stalks. At the worst spot — a bog hole that I could not go around and where I suspected Greg had once mired his Geo — I gunned the engine and plowed ahead.

Momentum carried the day. We shimmied free, and soon I could see Stacy and Shannon's truck in the center of the road with the tractor idling alongside. As we drove near, the big diesel throttled up. Belching black smoke, the machine resumed its work. It passed slowly, the driver flashing an openhanded salute from the cab.

Stacy and Shannon stood at the back of the pickup; and when I walked to them, leaving my parents to wait for a moment in the car, I could see that the rifle lay flat on the truck's bed beside a brown tarp and a camera. The two men were discussing something, and they stopped as I reached them.

The air hummed with a claustrophobic charge, and the cornfield seemed hot. It was a chore to breathe or talk.

"She's dead?" I asked after a moment.

"She is," Stacy said, speaking as low as ever, though with a tremor in his voice. He looked at the corn.

Shannon nodded, and as the three of us stood at the tailgate, I thought that grief was running below the surface in both men. They were professional, though, and busied themselves extricating tools from the packed cab. Speaking little, they moved with practiced efficiency.

Many questions occurred to me, but I settled for asking whether they needed help. Stacy startled when I spoke, looking up from a printed sheet of paper that he was filling out.

"We do, actually. We could use your help getting her out of the corn and onto the truck. But ask your mom not to take pictures, since there'll more than likely be a federal investigation into how this bear got shot the first time."

I beckoned to my parents, who emerged and walked slowly up the road. I left them beside the pickup, standing between walls of tall dry corn. Following Shannon, I found that we didn't have far to go. Twenty paces into the stand, we came to the place where the sow had fallen. Shannon handed me a pair of blue exam gloves.

"Don't let her touch your skin," he said. "We don't know what she might have."

He grimaced. "No bear should smell like that."

The stink hit me a moment later, filling my mouth and nose so I had to fight against retching. Decay hung heavy on the air, as though the bear had been dead for days instead of minutes. Everything in me wanted to recoil and run. At my feet, the sow lay with her paws and legs in the aspect of walking.

She was worse than shrunken, bringing to mind the way both my grandfathers had looked on their deathbeds. Her flesh, like theirs, had melted away, leaving skin loose to the touch and the body's bony armature showing clearly.

The sow lay in slatted shadows, and Stacy knelt to examine a small metal tag in her ear. Though only a scrap of steel embossed with numbers, it meant something to him.

"I know this bear," he said softly. "Her name was Millie."

Standing, he took the camera from its case.

"She was a good bear. Never got in trouble far as I know."

It was very still, and I listened to the field's small noises until Stacy shook his head and turned to business.

"Get her sternally," he told us. "I want a good picture of the head."

Working together, Shannon and I rolled her. As the sow's weight shifted, the three-inch curl of her claws brushed the back of my hand. They were chalked with mud; and crossing my fingers, they puckered the blue rubber gloves.

The smell got worse as we held her, and I could see that her sores had degenerated into outright rot. When the bear was properly arranged, Stacy snapped photos from many angles. I had time to study her closely, to look at the grievous wounds and sloughing flesh, and to understand how much she must have suffered. I tried, but failed, to see the glint of her eye amid swollen, suppurating tissue. Without moving my hands or her head too much, I felt in her fur for one of the thistle heads that matted her coat. Rolling the small spiked ball between my fingers, I looked at her teeth.

Her condition had not diminished those yellow plaque-rimed weapons, and I could not help thinking how easily the bear's mouth could enfold my arm. That was my enduring memory of the sow: In death, I better understood her strength. Laboring to support her neck and seeing Shannon doing the same on the far side, I was amazed at our frailty.

When the sow moved, her head dipping forward by an inch as if of its own accord, Shannon and I stiffened. I nearly asked whether he had jostled her but held my tongue. He stayed calm as a viscous stream of blood — red, bright, and wholly different than the drainage from her older wounds — poured from her nostrils. It fell to the ground and was swallowed, leaving a stain in the dirt.

Leaning forward, I craned my neck to see more. On her brow line, centered between the eyes, was a small dark hole still wet with blood.

"How?" I said out loud, without meaning to or wholly grasping what I meant by the question.

"Driver saw her," Stacy said, putting his camera away and unfolding a brown tarp on the ground.

"Saw her on his first pass, not far from where we had the trap set. When he came around again, I climbed up on the hood of the tractor. She was right alongside, stumbling like she could hardly keep her feet. Couldn't keep pace with the chopper, even, and when she came to a thin spot in the corn, I took the shot."

"The first shot," I said.

"The first one knocked her down, and I shot again from the tractor."

"And the third?"

"The third one" — he looked down at the bear — "was because I wanted to be sure."

Lifting the tarp from the corners, we bore her away. Except for the thick-boned bulk of her head, neck, and shoulders, the sow was surprisingly light. Her smell came up in sickening waves, and when my hands slipped lower once, liquid dripped from the tarp's bunched edge.

"We'll know soon," Stacy said when we reached the truck. "We'll know what killed her when we get the autopsy from the state veterinarian. But looking at it, I'd say bird shot to the face and then a hell of an infection. I'd say, too, that she kicked those cubs off when she got so weak or skinny she couldn't nurse them, and that's when she came to the field."

My father joined us at the bumper. He helped us lift Millie into the bed of the truck.

"Let me know," I said to Stacy. "Please let me know what you find."

He climbed into the driver's seat. Shannon wrapped the tarp tightly around the bear's carcass, hiding it almost completely. Because there was no space to turn a car, I drove out of the field in reverse with Stacy and Shannon following. Whenever I took my eyes off the rearview mirror, I saw their grim faces.

Leaving the stand, I backed a wide turn across the field's dirt margin. The truck passed and was quickly down the road. In the car, we sat and watched the chopper working a good way off to the south. The dump truck reached it, paused while the hopper tilted, then pulled away on its path to the dairy.

Following it out, we turned south and made for the highway. It was quiet in the car, and thinking of the sow, I decided that I had never seen a clearer case of mercy killing.

I WAS not long in getting angry. Who shot her to begin with, I wondered, and why? Someone had pulled a trigger, dooming Millie and her cubs. That knowledge set my hands tightening on the steering wheel until the plastic groaned.

The highway unwound, and just before we passed St. Ignatius, my mother spoke from the back seat.

"I will be thinking about that bear," she said, trailing off.

The road ran straight. The sun, yolk yellow as it always gets in September, painted the wayside grass.

"I'll be thinking about her," she said, "for a very long time."

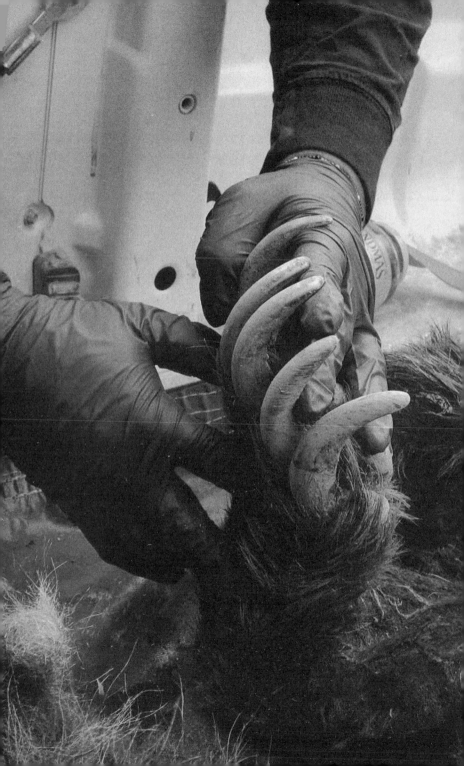

8

Fallow

I WALKED WITH Tick in the shorn field, crossing it from east to west to pick up the last trail cameras and set my fence right for winter. By late afternoon, the October sun had warmed the earth and a west wind rattled in the stubble. It was fickle and hard, and stronger gusts pried at the soil, which had dried and cracked since the pivot quit circling. The clouds were high; the light pale, diffuse, and hard on the eyes.

Wind fingered the earth, dislodging clods like a child picking at scabs. The dog trotted easily ahead, investigating tangles of stems and old bear sign. Dust rose in curtains, sliding eastward over the empty ground.

Reaching the west fence and slipping the cameras from their posts, I slung them across my shoulders, turned, and began walking with the wind. A red-tailed hawk crossed overhead with a white plastic sack soaring beside it. Bird and bag coasted together, fast, east, and away.

I made for a small patch of corn that had been left standing in the northeast corner of the field, a swath thirty yards wide and a hundred long. Ever since Greg had shut his cutting crew down, the patch had been a source of consternation to me.

"I've got plenty," he said when I asked him about it. "With that new fence and the good moisture this year, I've got more corn than I can pack in my silage pit. So, I figure I'll let the cows eat it in the field."

"What about the bears?" I asked, knowing that at least one grizzly remained in the stand.

Greg shrugged. "They don't bother grown cows."

At first, his stock had worked the margins of the stand, thinning the crop and trampling stems into a knee-high tangle. I had watched closely, thinking that cows wasted a good deal more corn than bears did and wondering if I might see a heifer jerked from view like a swimmer taken under by a shark. It never happened. The cattle ate for all they were worth while the stand frayed and thinned.

Crossing the field and pausing south of the remaining corn, I saw daylight peeping back from the north. Then I knew that the bears had quit the crop and field. No cover remained, no shadows in which to hide.

I found and knelt in a shallow daybed. Running my hands along the rim of the excavation, I wondered if Millie had dug the thing. Whether or not she or another bear had made it, the pit's architect was long gone. The cattle knew this as well as I did. They moved freely through the last stalks, pillaging like an army in a conquered town.

The field seemed lonely and pale, as if color and life had followed the bears to higher ground. Crouching in the bed, which was no more than fifty paces from where Millie met her end, I felt a bitter desire for retribution. The feeling had been with me since I had helped Stacy and Shannon carry the sow out. From the moment I had lifted my corner of the tarp, I wanted somebody to pay for the damage that had been done to her.

I could have forgiven an accident. If a bird hunter had run across Millie and fired a round in a state of terror, I would have understood it.

But the timing was wrong for a hunting accident. The reservation's earliest game-bird season — partridge — began on September 1. From the look of Millie's wounds and the advanced condition of her wasting in the images we captured of her, Stacy and I agreed that she had likely been wounded in mid- to late August — a time when nontribal sportsmen could not legally hunt the hedges and thickets near Schock's dairy. When I raised the prospect of a tribal bird hunter running into Millie — tribal members were not bound by hunting seasons and could take animals year-round — Stacy said that it wasn't likely. He told me that Indians were focused on big game in late summer and were more likely to kill birds, when they wanted them, with small-caliber rifles.

Gusts scudded over the field. Flocks of blackbirds gleaned the remaining kernels, and sometimes the movements of grazing cattle sent them up. The birds banked and split, reformed in a cloud, and plunged to earth.

They flew, turned, and settled several times. Then they rose higher, chattering. Thousands of wings caught and reflected the pale light, the flock becoming the color of the sky. When the birds slowed and faltered, the wind pelted them eastward as if they were handfuls of gravel.

A good-size dust devil chased them toward Millie's Woods. Leaving the hardpan, the whirlwind lost its substance from the bottom up. Dissipating above the grass, it vanished.

BRANDON COUTURE —a game warden for the Confederated Salish and Kootenai Tribes, and the person tasked with investigating Millie's shooting—came to meet me at the northeast corner of Schock's field. He wanted a copy of my footage of the sow and to compare notes about the timeline of her wounding and death. From the first, I was struck by the warden's serious mien and carriage. He isn't much taller than I am, but as we spoke, I couldn't help but feel as if I were looking up at him.

Brandon is a broad-shouldered, dark-haired, imposing man. Like Stacy's and Shannon's, his last name has history on the reservation. There is a Couture Loop just south and west of Arlee, and the surname appears frequently in the ranks of tribal employees. Like many people on the reservation, his ancestry is mixed. He told me once that because he's only a quarter Salish he "got it from both sides," often finding himself listening to whites bad-mouth Indians and vice versa.

Brandon's gear skewed military. He wore a bulletproof vest, a sidearm in a tactical-looking thigh holster, and dark fatigues that were a good deal more imposing than the collared shirt and blue jeans favored by Montana's state wardens.

We stood by his pickup and discussed the difficulty of tracking down Millie's shooter. Brandon told me all that he could about the investigation. Because he is a naturally effusive man with a tendency to let stories meander, our conversation stretched out to a half hour, detouring through a discussion of the illegal eagle-feather trade and a lesson on best practices for interviewing suspects.

Like most wardens I've known, Brandon takes his job seriously. He has an honest affection and respect for animals and a deep dislike for people who harm wild creatures beyond the limits of the law. As a hunter, he goes into the higher reaches of the Mission Range in search of mountain goats. If he approaches those hunts as he does the work of investigation, I have no doubt that he often gets his billy.

Brandon's task was a daunting one, and he knew it. Wildlife crimes are hard to solve, as poaching seldom happens in broad daylight or in front of witnesses. That the sow's wounding had happened in or near Millie's Woods — an area known for being difficult to police and dangerous to enter — didn't help matters.

Still, Brandon was hopeful. Grizzly bears are endangered throughout the Northern Continental Divide Ecosystem, and shooting one is a federal crime. Because of this, he had the help of a wildlife investigator from the United States Fish and Wildlife Service. Brandon described her as petite, intense, and devoted to the cause of tracking down Millie's shooter. He didn't share her name, perhaps because her work is more safely practiced in anonymity. Here, though, I'll call her Jen.

Jen's help, and the involvement of the USFWS, raised the stakes of the investigation. Shooting an endangered animal — a grizzly in the Mission Valley, for example — carries a penalty of up to ten thousand dollars and five years in what Brandon called a "hard-time, no-joke penitentiary."

Although Jen and the Feds carried the "big stick," the tribes were responsible for much of the investigation's on-the-ground labor. That work began with postmortem X-rays and a necropsy, from which Brandon learned that Millie had been hit with a load of number eight bird shot, likely fired from a twelve gauge.

The mark left by a shotgun blast speaks volumes about the situation in which it was fired. There are more than four hundred BBs in an average number eight load. When the trigger trips, they rattle along the tube, caroming off walls and one another, each on a slightly different trajectory.

The rate of their divergence is determined by choke — the structure of the barrel's utmost end — and shot size. Experimenting with my own gun, I learned that a paper target hit by similar shot at three feet shows a ragged quarter-size hole. At ten feet, the hole has expanded and is hazed at the edges with pinpricks left by individual BBs. Farther out, the damage expands into a devastating cloud.

Nobody gave me a precise count of the shot found in Millie's head and neck, but Brandon did say that it was "most of a load." It is safe, then, to say that she had several hundred pieces of lead in her hide and sinuses. The dense pattern and close grouping of those fragments suggested that Millie had been hit head-on from a distance of around twenty feet.

Millie's story — the gruesomeness of her maiming and the extent to which she wasted — bothered everyone who heard about it. As soon as the necropsy results were in, the tribes organized a reward for information leading to the shooter's arrest and conviction. The state of Montana partnered with them to double the money, so the reward totaled five thousand dollars.

That kind of money tests loyalties, particularly in a valley where many people live below the poverty line. Sometimes a reward turns up useful information. Occasionally, it prompts enough talk to shed light on unrelated crimes. Without exception, it also generates a wide variety of nonsense calls.

When I first talked with Brandon, he was in the midst of wading through a morass of somewhat useful reports and suspicions. A few leads struck him as serious enough to follow. He pulled away from the field like a man on a mission, and as I watched his pickup shrink into the distance, I thought it likely that he would solve the thing before October was through.

AFTER BRANDON departed, I was gripped by a strange desolation. The warden had leads to chase. Hunting poachers gave him purpose. When bears were trying to get into the corn and my fence stood in their way, I had been similarly animated. But with the crop cut and the wild things gone, I lacked direction.

I stood at the field's edge. It was hardly past noon, but I was already finished with the day's work. Looking toward the mountains, not knowing what else to do, I decided to take a hike.

Driving up Red Horn Road with a thin rain patterning the windshield, I knew where I meant to go. Somewhere beyond the cusp of the foothills were the Ashley Lakes, a pair of pretty tarns.

The fringe of Millie's Woods ran close to the right-of-way fence. Signs reading CLOSED DUE TO EXTREMELY HIGH LEVELS OF GRIZZLY BEAR ACTIVITY and bearing the seal of the tribes hung from the barbwire. I drove onward past the upper edge of the closure to where the country road steepened and diminished to a muddy track.

Aspens closed in, their branches scraping the cab as I switched into four-wheel drive, then compound low. Leaves glowed gold. All had turned and some had fallen, making the ground and forest the same rare color.

When the truck could go no farther, I hung bear spray and a revolver on my belt and walked up switchbacks on an overgrown logging road toward a ridge. The forest was thicker than I had expected, with an understory that made it hard to see far in any direction. Old cedars grew in the draws where springs ran close to the surface. I crushed needles for their smell.

At one tight corner, a mound of scat filled one of the road's ruts. After stepping over it, I walked slower. Beyond the ridge, a steep-sided canyon ran back toward McDonald Peak. Descending into that cleft, I crossed from the foothills into the true mountains. The path was unmaintained and scattered with deadfall.

Down in the bottom, pines and Douglas firs gave way to something more like jungle, with the track wending through rain-soaked alders. I walked fifty yards farther before something in me rebelled. The thickets seemed heavy with the presence of bears, as the cornfield had been before it was cut. I hesitated for a minute, listening to an unseen creek. Turning, I abandoned the notion of climbing to the Ashley Lakes.

That grizzly feeling lingered in my mind as I walked out of the mountains. When a bull elk leaped up and sprinted through the forest, my hand went to the gun on my hip. The reflex made me mad. *Scared,* I reproached myself, *like the one who shot her.*

I was hungry when I reached the valley floor. Driving out, I stopped beside an apple tree growing on the edge of Millie's Woods. It was a gnarled thing, an unpruned-since-the-seventies relic of an early homestead. Though it stood just a little way from the road, I still felt like prey as I clambered through a barbwire fence and walked to the trunk.

The tree was plucked clean to a height of ten feet, with not an apple to be reached from the ground. Up higher, the branches sagged under the weight of red McIntoshes.

I climbed, gripping the trunk with my knees and using lower limbs as handholds. Ascending, I found myself considering Millie's cubs and wanting badly to do something for them. I didn't know what, if anything, might be useful.

Balancing in a fork of the tree, I snatched at fruit until two came away. Eating one on the spot, I decided to call the facility where the cubs were being held. It would be a start, at least, to ask about the orphans and learn whether it looked likely that they'd find a home. The idea satisfied me, and slipping the other apple into my pocket, I slid to the ground.

9

Visiting

O N   O N E   of the season's first frigid days, I drove to Helena. The Montana Fish, Wildlife & Parks' WILD Wildlife Rehabilitation Center did not strike me as a likely place for keeping grizzlies. Located just off of Highway 12 and bounded by a suburban park, a pond, storage units, and sprawling neighborhoods, the facility occupies around five acres. Half that land is taken up by a wildlife museum with a pretty stone facade and neat landscaping, which offers hunter-education courses and elementary school tours. The pens are around back, and as I parked in the large empty lot, I could not see into them.

I took the apple that I had picked in Millie's Woods from the door's side panel and slipped it into my jacket pocket. Slinging a camera over my shoulder, I walked into the museum and told the docent that I had come to talk with the woman in charge of the cubs.

I studied a full-body mount of a wolf while he spoke softly into a phone. "You can head back through the gates," he eventually said to me.

A scant layer of snow covered the ground, and I hunched against the wind that scours Helena in late fall. Focused on the prospect of seeing the cubs, I nearly had my hand on the chain-link fence —

which was ten feet tall and plastered with no-entry signs — before I noticed an enclosure to my left. Inside was a sun-whitened cottonwood snag with a woebegone bald eagle perched in its uppermost fork. The bird swiveled its head, tracking my movements as I worked the latch and stepped inside.

Lisa, the facility's wildlife rehabilitation manager, had been brusque when we had spoken on the phone. She was no less direct in person, though the way she discussed the cubs betrayed a streak of kindness.

A thin woman with graying hair and piercing eyes, she took my measure as we settled into chairs in her cramped office. To get there, we had walked through a long room past rows of empty cages. The place smelled of disinfectant, and from far off, I heard sounds of skittering movement.

"I've already exceeded my mandate," Lisa said. "Four weeks was my limit, and we're coming up on two months since those cubs came in.

"We're responsible for the care of these bears. We feed them, we keep them healthy, and if it comes to it, we put them down. But the United States Fish and Wildlife Service is responsible for signing off on where we can and can't send them. The Feds have their own rules about where bears can go. As of now, they require an Association of Zoos & Aquariums certification before any facility can take a grizzly. That rules out a lot of places that might be interested, and all the private sanctuaries, too. It also means that when no AZA zoo steps up to take an animal, like is happening with the cubs out there, we get stuck in limbo."

She glanced at a queue of emails on her computer screen, then back toward me.

"There's a lot of testosterone sloshing around at FWP headquarters, and people up there generally fall into three camps. You've got some people saying, 'Kill 'em,' others saying, 'Save 'em,' and a bunch of people saying, 'Well, they're zeroes in the population now, so it doesn't matter.'"

I asked what category she fell into.

"I don't want them killed. I've only put down one grizzly cub in five years, and that one came in hurt. No — I won't see them killed if there's any way around it.

"Last time we had a pair like these," she said, her voice dropping conspiratorially, "three places would've taken them, but none were AZA certified. They were ZAA certified, which is pretty much the same thing, only through the Zoological Association of America, but the U.S. Fish and Wildlife Service wasn't having it."

She described how close those cubs had come to being put down before a suitable zoo had been found, how hard it had been to find an institution willing to take on a pair of wild young grizzlies.

"Zoos aren't doing too well these days," she said, explaining that attendance and funding have flagged as the public grows less comfortable seeing animals in cages. "And bears live a long time. We're talking about a twenty-year commitment, a handler, and funding to feed them."

Keeping a grizzly confined is an expensive proposition. WILD, she said, was in need of a new bear pen because their two existing enclosures tended to overflow with inmates each autumn. The planned construction was relatively small-scale, a square measuring about fifty feet per side.

"You know what that's going to cost? Three hundred thousand dollars at least, and it'll take a year to build. Grizzlies can dig like you wouldn't believe, and we have to go down four feet with the footings. Even so, when I have bears in here, I do not have a comfort level. The *last* thing we need is to have grizzly bears loose in Helena."

A rehabilitation and holding pen, she went on, could be constructed for pocket change compared to a zoo exhibit. The cost of a permanent bear facility runs shockingly high. One built recently in St. Louis cost eleven million dollars, a number that does not include the salaries of the keepers and the expense of keeping a hungry grizzly fed.

Pushing back her chair, Lisa stood.

"We might as well try to see them," she said. "But don't be let down if they keep out of sight and you never get a picture. The guy from the local paper was here for hours before the cubs came out from behind their log."

We walked out across gravel into an alleyway between high chain-link fences. The cubs' enclosure was larger than I had expected, with a conglomeration of peeled logs and tractor tires at its center. It looked like an empty playground surrounded by a very high chain-link fence. Timbers stood at angles, and the vertices where they met were secured with chains and long bolts. Some of the tires and smaller logs were rigged to swing or pivot, and I wondered if captive bears ever relaxed enough to use them. On the space's far side, I could see a small concrete-banked pool of water.

Haggard grass tufted the ground, and halved apples were strewn all over. I was looking at the fruit when a flash of brown fur showed at the pen's far corner. A horizontal log blocked most of my view, but across it, I could make out the line of a grizzly's hump.

"Keep quiet," Lisa whispered. "They hear everything."

Ears bobbed in and out of sight, moving slowly toward the end of the log. Each time I raised my camera and snapped a photo, they stopped short and kept still for a moment before going on.

When the bear stepped fully into view, she was beautiful, with a thick coat and blond fur curling like a stole around the base of her neck. The cub, a word that seemed less fitting than it had in summer, was a perfect miniature of an adult grizzly: ruffed hump, dished face, small dark searching eyes. Though no more than two feet tall at the shoulder, she had acquired the solidity and gravitas of her species.

The bear chose her path carefully, looking often in our direction. She took three steps before another shape showed above the log. Although colored like her sibling, the second cub was more delicately formed and less sure of herself. Crossing into the open with halting steps, she kept tightly to her sister's flank until they both disappeared behind a jumble of tires.

I studied the pen, noting the electrified wires offset from the base of the chain-link, and the squat concrete building with guillotine doors that adjoined the enclosure on one side. The doors reminded me of the ones on a culvert trap.

Emerging again, the cubs stayed close to each other and moved like a single beast. Coming to the enclosure's pond, they walked its rim with their shape reflected in the water. After a summer of working to keep bears outside a fence, it was strange looking in at them. That they would never again go free struck me as a hard thing, and I wondered whether a caged life was better or worse than a quick, painless end.

A breeze hummed in the chain-link, and judging it loud enough to mask the shutter's noise, I raised my camera. Without a sow for scale, the cubs looked like adult bears except for their small feet and oddly fluffy coats. The rising wind played in their long fur, setting their silhouettes in motion.

Leaving the pool and stepping behind a long horizontal log, the larger one turned to face me, put one paw, then the other, onto the wood, and settled forward. Her sister did the same, and for a time, the two bears lay side by side with their forefeet across the trunk and shoulders nearly touching. They stared straight at me, scenting the wind. When southbound geese passed over in a racket of noise, the cubs watched the birds. I photographed through the chain-link until my fingers stung with cold and my breath frosted the view-finder.

The longer I studied them, the smaller the cubs looked. They shrank, the pen loomed large, and I twisted the lens's zoom until it seemed that we three were close together. I could see the grain of their coats and the whites of their eyes when they glanced at each other.

I documented the cubs with care. It made me think of my mother and theirs. When the camera ran out of memory, I balled my hands and stuffed them into my pockets. Curling numb fingers around the apple from Millie's Woods, I studied the detritus on the ground within the pen. Halved fruit lay everywhere in the dirt.

"Where do you get their food?" I asked softly.

"A grocery store gives us what they can't sell."

The apples were Honeycrisps well past their prime, and look-ing at the mottled skins and browning flesh, I could imagine their mealy sweetness.

"I wish they'd eat more," Lisa said.

I drew out the McIntosh. It was an old-strain fruit, as hard and good as western Montana soil. Telling her that it was from the Mis-sion Valley, I asked permission to leave it.

Rising, she led me toward the shed that formed one wall of the pen. The cubs drew back when we moved, watching closely until we were out of sight. Stopping, Lisa gestured to a small steel door in the wall. Sliding back the bolt, I pitched the apple through the gap and sent it bouncing across the cement floor.

As I set out homeward in a building snowstorm, leaving that apple seemed like pure sentimentality. But while Helena faded be-hind me and the road climbed, I let myself imagine the cubs finding my gift in the concrete den where Lisa said they slept. They would sniff it with magnificent noses, and sorting through my unfamiliar scent, they might find some hint of the woods, the valley, and home.

With the engine wound up and a flurry beating past the windows, I wondered what that could mean to a bear. I hoped that a familiar taste might offer a measure of consolation.

A plow worked the lane ahead of me, pitching a white wave into the ditch. Downshifting, I considered Lisa's final words.

"I don't know what you can do to help," she had said, "except get the word out. A famous bear has a better chance than one nobody knows about. Famous bears are harder to kill."

LISA'S COUNTERPART at the United States Fish and Wildlife Service was a man named Wayne Kasworm. They had already reached out to the AZA's Bear Taxon Advisory Group. Since their appeal to that body of zoo directors and curators had yielded nothing but a handful of regretful denials, I decided to look around on my own.

I started with the nearest and most likely options, writing emails to ZooMontana in Billings and the Grizzly & Wolf Discovery Center in West Yellowstone. Both institutions responded quickly with a good deal of concern for the welfare of the cubs. Both facilities were also chock-full of bears.

I remember a call with an employee of the Discovery Center particularly well. He seemed to honestly care about what happened to the cubs. But he was in the midst of fund-raising for a new permanent facility and could take in no new animals for several years. We lamented the timing together.

Afterward, I sat by the phone and wondered what the hell to do. I thought about Lisa's advice about "famous" bears and wrote something intended to raise the awareness of the cubs. It ran prominently on the home page of the *Daily Beast* under the headline — theirs, not mine — WON'T YOU PLEASE ADOPT A GRIZZLY BEAR CUB?

In sending the missive, I hoped that someone might present the issue to a zoo director capable of doing something and that an outpouring of care and concern for the cubs might change the trajectory of their lives.

Responses poured in. The good-hearted staff of People and Carnivores collected donations to help transport and house the cubs if a zoo could be found to take them. Although we didn't raise gobs of money — all told, a couple thousand dollars arrived — the effort attracted some attention as well as a good deal of advice about what should be done with the cubs.

In the end, though, nothing came of it. I went on calling zoos and sanctuaries for weeks, reprising the same heartfelt, exhausting conversation. After a while, I could recite its crucial lines:

"I wish you luck," the zookeeper would tell me. "But we don't have room."

"We've raised money," I'd say.

The keepers seldom asked how much. Instead, they explained that taking in a bear was too much of a commitment and that zoos were more interested in exotic ursids — sloth bears and sun bears — than orphaned grizzlies.

The calls wore me down. While talking about the cubs, I saw them as clearly as if they were in the room. Their image was slow to fade, persisting until it seemed the bears had taken up residence in my mind and might den there through winter. When weather swept through Missoula, I pictured it reaching them in Helena. The cubs shivered in bitter gusts. They panted through the season's final warm afternoons and dripped rainwater when Chinook winds blew from the southwest. The cubs never appeared without a backdrop of chain-link fence. I did not like it much and grew desperate to get them out of the cage and my head.

OCTOBER'S END was troubled and strange. With every passing day, I grew more conscious of the strain that Lisa was under to dispose of the cubs. She cared about the little bears and would save them for as long as she could, but I knew that she couldn't keep them forever.

I called a series of less and less likely institutions. The cubs were still in Helena come November, the month when the cold bites hard. Quick, mean blizzards blew through in succession, tearing branches from trees and whitening the ground, howling in the chain-link. Throughout those weeks, I thought about the bears with dismay. Having done all that I could think of on their behalf, nothing remained but to wait.

The days shortened, and snow piled higher in the mountains. Winter closed around Missoula like a fist, with the last heat and light of autumn showing faintly through its fingers. Windows glowed, and the stars were bright and clear.

I never forgot the cubs. They came to mind each day. But something began changing. When I thought of the sisters, I could no longer picture them clearly, see wind lifting their fur against its grain, or conjure the fear and curiosity in their eyes. I did not cease caring, but the cubs were further from me.

It went like that, with hope guttering like a candle, until I ran out of zookeepers to call. The thing seemed finished and Millie's daughters as good as dead.

10

Hunters

W HILE I worked and worried on behalf of the cubs, Brandon kept on with his investigation. From the time that the reward was mentioned in the local papers, he had listened to a steady stream of stories. A farmer had seen magpies and, later that day, a couple of grizzly cubs. "Case closed," the guy told Brandon. "You can send a check." Somebody else knew beyond a shadow of a doubt that Millie had been shot by a pair of neighbor kids who liked to spotlight deer at night.

Tips came in from all over the reservation, but Brandon didn't believe that a gravely wounded bear would travel far. He went on listening, waiting for someone to mention Post Creek or Millie's Woods.

On the sixth of October, Stacy put Brandon in touch with a woman who lived near the woods and claimed to have information about the case. "You need to talk to her," Brandon remembers Stacy saying. "This sounds legit."

From the first time he spoke to Laura on the phone, Brandon thought that his witness was on the level, if somewhat odd.

"She was one of those off-the-grid types, real self-sufficient. She didn't know about the reward as far as I could tell, and I never mentioned it to her until after we had spoken in person."

He said this of her while we were driving north in his pickup truck from the town of St. Ignatius. We were en route to the area around Millie's Woods so Brandon could bring me up to speed on his investigation.

His first interview with Laura was conducted at her house, which had a commanding view of the woods. It left Brandon feeling like his case had broken wide open.

"About a month ago," she told him, "I was working on my roof and heard somebody shooting."

Peering in the direction of the noise, she saw two small brown forms—"two humps"—burst from the brush and run pell-mell through the tall grass along the bank of an old canal. She recognized the cubs by their shape and color because she had seen them before, always with a sow.

That day, though, no mother appeared. As the cubs rushed headlong through the grass, a man stepped from the fringe of the timber. He was tall, as she recalled, and thin. Holding a shotgun in his hand, he fired at the cubs as he walked. There might have been six or eight shots, she told Brandon.

It is a long way from the corner of her roof to where she said the man came into sight—several hundred yards. She yelled and cursed across that distance because she loves bears and hates poachers. It must have taken courage to shout, as exposed as she was, and with him blazing away. Laura screamed, but he took no notice or didn't hear. Firing a final round, he turned and disappeared into the woods.

Laura told the story in detail, tracing the route that the cubs had taken in their flight. She had seen the man clearly, too, and described his hair, glasses, and manner. The way she talked gave Brandon confidence.

When Laura said that she thought she knew who the shooter was and that he lived in a nearby house, the warden's hopes soared. She had even spoken to the guy once while looking for a lost dog. He had met her in the doorway, and she remembered him as quite tall and a bit, as she put it, "slow."

At certain points in the telling, Brandon said, Laura really worked herself up. Having spoken with her myself, I know what he means. She has a habit of darting headlong toward outrage, with her voice rising as she recounts injustices.

Fortunately, Laura calms down as quickly as she winds up. When calm, she makes a lot of sense. Brandon thought so, too, and came away from their first conversation with the impression that she was well-intentioned and honest, if a bit eccentric.

We parked on the corner where Red Horn Road bends south and becomes Foothill Road, and Brandon explained how he had thought highly enough of her testimony to interview other neighbors around Millie's Woods. None had seen the shooting, but several of them had suspicions about the people living in a pair of houses on the edge of Millie's Woods — the place from which Laura's gunman had emerged.

The houses, both owned by the same family and located on a tribal land lease, were described to Brandon as a source of consistent mischief and trouble. The neighbors' stories centered around a tall, gray-haired father and his two adult sons. The elder son was short and stocky. The younger one, who Laura claimed to have seen shooting at the cubs, was lanky and bespectacled.

As soon as Brandon started talking to the neighbors, he was plunged into a local history of poaching and trespass. One neighbor believed that the younger son had stolen a motorcycle from his place. Another accused the elder son of vandalism. None of this is rare in rural Montana, but the number of reports and the fact that all three men had served time for nonviolent felonies — theft, drug charges, and drunk driving among them — were enough to draw Brandon's attention.

Pulling the younger son's mug shot, Brandon arranged a photo lineup. Laura picked out her gunman from a group of six strangers without a moment's pause. "That's the guy," she said, in spite of the fact that the photo was years old. "Except now he has bangs and glasses."

Brandon still remembers that lineup because he was struck by how well it went. "There wasn't any hesitation," he said. Laura looked at the photo, pointed the guy out, and that was that. She was, he told me, certain.

Settling back in the driver's seat, he looked out across the valley. From where the truck was parked, we could see most of Millie's Woods. The aspens had dropped their leaves, and snow covered the ground. With most of the branches bare, I could see into and across the grove. In that monochrome landscape, it was easy to pick out the shapes of the few houses that nestled close to the woods. Smoke rose from one of the nearer chimneys.

Brandon visited the suspects' house soon after the lineup, along with the older son's probation officer. Coming up the driveway, he caught the scent of rotting meat.

"There were deer and elk hides everywhere," he said. "On the porch, hanging over the railing. Pretty fresh hides, all drying out like they were going to use them for something. That's normal around here. A lot of our families use hides for traditional stuff. They'll brain tan or whatever."

He wouldn't have thought twice about the hides if it hadn't been for the proximity of Millie's Woods. The outermost edge of the grove was just a hundred yards from the house, and Brandon knew that if he could smell the hides, bears would be drawn to them from all over.

Pretty big attractant, he thought as he walked past fruit trees and bear scat in the yard. Entering the house — with a resident on probation, no search warrant was necessary — he began interviewing the men while the other officer searched for firearms. They turned up marijuana, crushed pills, and more than one weapon but no shotgun. There were also a suspicious number of eagle parts around. Though tribal members are allowed to keep raptor feathers when they find them, whole birds are supposed to be turned in. The father claimed to have found a bald eagle's carcass, but Brandon had his doubts. The younger son, the primary suspect, was at work that day and did not come home.

It all seemed shady enough that Brandon wanted to search the other house, which stood just a short distance off, but since the probationer didn't live there, entering the place would have required a warrant.

"I'm not saying that people who go to prison never change," Brandon told me. "A lot of people, they make mistakes and get better. But some of them are career criminals, and they're good when it comes to interviews. They know just what to say, and what not to say."

That was the feeling he got from the father and elder son. Brandon started with simple questions and found both men suspiciously closemouthed. He asked whether he'd seen grizzly bears around, and the father replied: "No. Never."

"That's what he told me," Brandon said, showing his teeth in a tight, incredulous smile, "while we were sitting next to Millie's

Woods, where everyone around here knows you come to see bears. Where you can't walk fifty feet without stepping in scat."

Pressing his suspect, Brandon asked the man if he had ever hunted the woods and whether somebody might have seen him in the trees with a weapon.

"I never hunt the place," the father said, though, as members of the tribe, he and his family had every right to do it.

"What about those hides?" Brandon asked. They went on like that for a long while, making little progress. After several rounds and strong words about obstruction of justice, Brandon managed to establish only that both men occasionally hunted deer and elk and that they had seen a grizzly at least once or twice.

He left the house stymied, having learned almost nothing and having missed a chance to talk with the younger son. Walking out, he passed the hides. "Tons of them," he told me. "And the place where they were stored wasn't secure. Anything could have come in and got them."

PUTTING THE truck in gear — he had left it idling while we talked — Brandon headed down Red Horn Road toward Schock's cornfield. As we cruised along the gravel, he described how things had gone when he and Jen, his counterpart at the U.S. Fish and Wildlife Service, brought the younger son in for questioning.

"For the most part, you can tell when people are being honest," he said, shaking his head like the whole thing still perplexed him. "He seemed like a nice kid to me — I say kid, but he wasn't a juvenile. Night and day from his brother and father. Didn't seem to have a mean bone in him. Like Laura said, he was a little — simple."

As far as Brandon could tell, his suspect was straightforward and honest, holding nothing back. Without meaning to, he incriminated a family member in an unrelated wildlife infraction. After the interview, when the younger son had left the room, Brandon turned to Jen, and said, "There's no way that guy's the shooter."

The younger son struck both investigators as a person unlikely to blast a grizzly in the face, but Brandon could easily imagine the father and brother pulling the trigger. His suspicions sent him and Jen back to Laura's place in the hope of filling gaps and moving the stalled case forward.

They showed up hoping for clarity. In short order, though, the conversation became a debacle.

"The details got muddied," he said. As he tried to lock in the particulars of what Laura had seen, her story began changing. He first noticed the inconsistency when she demonstrated the motions of shooting and reloading in a way that would have worked for a lever-action rifle, not for a shotgun. Then, when pressed, she couldn't remember if the man had been firing from the shoulder or hip.

The discrepancy seemed significant to Brandon. Shooting from the shoulder suggests taking careful aim. Hunters, whether after birds or big game, fire that way when they mean to kill. Shooting from the hip is different. Unaimed, quick, and offhand, it is the sort of thing that might be done to scare or drive away a creature.

Brandon tried to dig deeper, asking specific questions, but things went from bad to worse. For every hole that he managed to fill in the narrative, two new ones emerged.

"All the details were different. All of a sudden, the shooter is more than six-feet-six. I'm six-feet-three, and I know that every one of the suspects is under six feet. After we did the interview, I was kicking myself. She wanted to help so much that I think she started making up information — trying to fill in the blanks with what she thought I'd want to hear."

I've talked enough with Laura to understand how difficult it can be to elicit a clear story from her. This is partly because of her natural strangeness and partly because of her circumstances and surroundings. Laura is always on edge and has good reason to be. She has had livestock killed, she believes, by both bears and neighbors. She has been bitten by free-roaming dogs — animals that she eventually locked in her barn and shot dead — and she's been harassed by hunters and threatened by cops. She claims to have evidence that some of her neighbors, including the suspects, are cooking methamphetamine in Millie's Woods. Laura packs a gun, and if half of what she says is true, I don't blame her.

In spite of the inconsistencies and digressions, Brandon never doubted the basics of her story — that she had seen a man emerge from the brush and fire at the cubs. Hoping to corroborate her account with physical evidence, he and Jen decided to search the woods. Finding even a single cartridge would have been enough to keep things moving and refocus his efforts.

BRAKING TO a stop on the gravel road, Brandon eased his truck through a three-point turn and started back for St. Ignatius. Glancing out his window at winter-killed grass and leafless aspens, he told me how he and Jen had walked down from Laura's house into the area above Millie's Woods. That space, which Laura calls the "corridor" because it is such a highway for bears and other creatures that come out of the mountains, is relatively open and looks a bit like a savannah. Clumps and veins of hawthorn and brush mark the places where water runs. Scattered wild apple trees cast shadows on high grass.

Fresh grizzly scat lay everywhere, but Brandon couldn't find a single shotgun shell or rifle casing. The operation was less of a needle-in-a-haystack search than it might have been because Laura had pointed out precisely where the man had stood when he had fired.

Brandon looked hard. He inspected the trunks and lower branches of trees, hoping to find pellets embedded in the bark. He and Jen spent hours searching. Coming up empty-handed in the places Laura had indicated, they moved downhill into thicker brush and scoured every game trail they could find. It was a fine idea. In Millie's Woods, such paths are the only good way of getting around.

Leaving the trails, they crisscrossed the margin of the woods. Hawthorn and head-high wild roses pricked them. The grass grew tall and thick enough that parting it was like wading through water.

"*Nothing,*" Brandon said, slapping the dash. "And the witness said that she had heard a bunch of shots. Between the two of us, we should have found at least one shell."

He told me, too, how it felt to wander around Millie's Woods while the apple trees still held old fruit, with bear scat underfoot. He was right to be nervous. Not a month before he made his search, and no more than five miles away, a bird hunter had run afoul of another grizzly sow and cubs. Brandon, privy to the details of that case, knew how the man had stumbled onto the bear in high autumn-browned grass; how the sow, rising enraged from her daybed, had charged him from a distance of ten paces; how in the course of falling backward, the man had shot at the bear and missed her cleanly; how she had bitten him once on the booted foot before gathering her cubs and slipping away.

That story was fresh in Brandon's mind as he searched for the break his case needed. From working with Stacy and Shannon, he also knew how many bears used the woods. Sometimes, when things got particularly bad at Schock's cornfield, the biologists asked him and other wardens to bring deer carcasses to the upper end of the woods. The idea was to draw the bears uphill away from the crop, and it worked well enough that the whitetails disappeared overnight.

"I remember dropping one off, and the next day, it was gone. Then I dropped four off, and by morning, all those were gone, too. So I knew how much they were using it. Walking through, you spook up pheasants and Hungarian partridge. Everything that moves in that brush, you think it's a bear. You're on edge, looking over your shoulder, thinking how if you pop through brush like this and run into Mom, she's going to teach you a lesson."

With such thoughts running through Brandon's mind, he and Jen continued their search into the shadows of taller trees down toward the suspects' house. But for all their labor and risk, they found nothing.

It was on the uphill walk, heading away from the woods back toward Laura's house, that Brandon lost hope of solving the case.

"We were so close," he said to me as we turned onto Hillside Road. The engine spooled up, and hot air blew through the vents. Now and again, the radio beeped and crackled with news. Elsewhere on the reservation, a deputy was "making contact." Somewhere on the highway, a semitruck had broken down.

"So many things could have gone differently: if I could have caught the one on probation with a weapon; if I could have found a shell down there by the woods; if I could have cracked the father when I interviewed him — got him to lose his cool.

"*Everything* pointed toward that house," he said with a fierce shake of his head. Another neighbor — not Laura — had told him a story about the time his dog had stolen a hide from one of the fence posts and brought the trophy home. The father showed up hot on the creature's heels, shotgun in hand, saying, "If I see your fucking dog again, I'll fucking shoot it."

Brandon drove faster, and we passed houses more frequently. Soon we were looking through the cracked windshield at the sporadic traffic of a St. Ignatius afternoon.

"But I couldn't make it add up to anything," he said. "Nothing that would stand up in court."

Pulling into the lot where we had met earlier in the day, he parked and idled. Gathering my things, I posed a final question.

"How do you think it happened?" I asked him. "Leaving aside what you can and can't prove, how do you think she was shot?"

Brandon took some time to answer.

"If the way you make a living for your family is tanning hides or making drums for people," he said finally. "If that's how you make your money, and bears are coming in and taking it right out of your pocket." He let another quiet moment pass. "The cubs. Maybe the sow knew better. Maybe she'd had an encounter with them before. But those cubs didn't, and I think maybe they were going up there, pulling hides off, and he went out with a shotgun. You know mama bears aren't going to let anything happen to their cubs, so maybe she confronts him, charges him, and he takes a shot.

"Something like that, he could have called it in and said, 'She charged me, and I dumped her.' That wouldn't have been criminal."

He shook his head and slapped his hand heavily on the steering wheel. "But after you blast the sow, why shoot the cubs? They got peppered, too, and while they were running away. The Feds looked at all that. She was shot from twenty feet, based on how tightly the pellets were grouped in her face and how many were in there. It was almost the full load. I don't see how something like that would hit her cubs, too, unless they were close by her side and facing the other way.

"You know this," Brandon said, looking me in the eye. "But when Stacy caught up to her, she was blind. Her nasal passages were split open. It was sickening to see. I mean, I've seen — being a hunter and a warden, I've killed a lot of animals. I've had to put down animals that were injured and maimed in terrible ways. But when I saw her, it was heart-wrenching. How mangled she was by the shot and how long she had lived like that — she lost hundreds of pounds. It didn't sit right with me. None of those others affected me like that, but seeing her was different."

Brandon stopped talking, and I cracked the passenger door. I was about to thank him for his time when he spoke up again.

"I would almost say that I got emotionally attached. I wanted to get those guys in a big way. I was upset, and I'm still upset. I really wish — of any case, I wish that this one had broken, and we had caught them."

11

Millie's Place

On a short overcast day near the winter solstice, I went into Millie's Woods. It was damp and cold, and blue-gray mist trickled down from the mountains, making the distant trees look dim. Snow had fallen recently, and though wind had freed the open, grassy places, an ankle-thick layer remained in the grove.

I balled my hands against the chill, clenching and releasing my fingers to drive blood to their ends. Approaching from the north, I followed a two-track road through a heavily grazed pasture toward a pair of trailer homes.

As I neared the first house, a skinny tortoiseshell cat slunk from the grass and jumped the ruts. Another cat slipped into view, then several more. All were similarly colored and certainly kin, and the field crawled with them. I kept on, with cats popping into view. More than a dozen of them ran soundlessly away, disappearing into the single-wide's cracked skirting.

The cats gave the place a surreal, wonderland feel, which was increased by the presence at the roadside of huge wild rosebushes — vicious, pretty plants with leafless stems the width of my thumb, hips like red Ping-Pong balls and thorns three-quarters of an inch long. Plants grow exceptionally well near Millie's Woods, as if nourished by the grove's wildness. The roses are common and can be ten feet tall. Other shrubs thrive similarly. In the sloughs, cattails reach higher than I can.

With the aspen leaves fallen for winter, the color of the pines —a green deep enough to be black in flat light—predominated. A gate of welded metal tubing blocked the road, and stepping around it, I followed the wheel tracks underneath the first branches.

As far as I knew, the bears were gone. Stacy's telemetry data showed that all of his collared grizzlies had denned in the mountains. Still, I was careful and stopped often to listen. The silence was pregnant, and it hissed in my ears. Even without bears, the woods kept their foreboding character.

The road bent left uphill. I walked it slowly toward the homestead where Millie Morin, the woman for whom the woods were named, lived for seventy years. People remember Millie, though she died in 2008 at the age of ninety-five, because she was resourceful enough to thrive on a forty-acre farm at the edge of the woods. She outlived four husbands and generations of bears there. Millie milked cows and somehow kept their calves from being eaten. She excavated ponds to catch the runoff that trickles out of the woods and raised trout. Nobody knows how she got along so well with the grizzlies. When Chuck Jonkel, a renowned bear biologist, asked about it, Millie pointed to the three-foot-high split-rail fence at the edge of her yard.

"They stay on that side of the fence," she said, "and I stay on this side."

By all accounts, she was a force to be reckoned with. Remembered by her son in an obituary as "not mean tough, but tough," she kept a close and uncompromising watch over the land around her home. She loved bears, sometimes carried a gun, and liked to run off poachers. When Stacy talked of her, he described a woman not to be crossed. He said that she "kept an eye on things," and nothing moved through the woods without her knowing it. From him and others, I formed an impression of a fearsome sprite — a more-than-human being who had absorbed the grove's character through long exposure.

Her house is gone now — demolished by a habitat restoration crew after the tribes bought her farm. The ponds and orchard trees remain, and an echo of care, so the clearing feels tended. Stopping in its middle, I found the silence broken by the noise of moving water.

In the ten years since Millie passed, the woods have closed around her efforts. Branches reach in from all sides. An old spruce — which must once have been a backyard planting — stands out from other growth. That tree, which belongs higher in the Mission Range, looks strange among pines, aspens, and gnarled apples. In 2006, when he first caught the sow and named her for the old woman of the woods, Stacy had anchored his snare to this spruce.

Trees are slow to forget. Bones, in our climate, are a long time in rotting. Still, I crossed the clearing and found the trunk abraded to a height of six feet, its bark rasped back to reddish younger layers. The bones of midsize animals — deer, sheep, or goats — lay scattered around its base.

I don't know how old those bones were. Stacy might have baited and trapped a bear in the same spot more recently. But the tree's look and the place's inhabited feel made it seem as if both the sow and her namesake had only lately departed. With the sound of water, the clearing seemed calm and sheltered.

It is no wonder that Millie Morin loved the place and guarded it jealously. Lushness prevailed, and the silence left me unsure if time was passing. Game trails emerged from the woods in many places, making for the ponds and old apple trees, seeming to converge from several directions. But in spite of its wildness and undertone of danger, the clearing was sweet. I could see how a fearless person might make her life there as long as she did not for a moment lose sympathy and respect for bears.

IF BRANDON had come to grips with the fact that his case would remain unsolved, I had not. Still hoping that something would crack or some new evidence come to light, I set off through the woods on a trail, striking uphill eastward from the clearing.

The ground was unfrozen and sodden beneath the snow, and I had to cross and recross one of the small black-water streams that leak out of the woods. It was difficult going, and my winter boots collected freezing mud. Brush grew beside the water, and I pushed through branches, sometimes snagging my coat.

Above the thicker woods, I walked south to the place where Laura claimed to have seen her neighbor shooting at the cubs. It was very pretty there, with scattered trees and the ground melted or blown clear. The mist had lifted somewhat, so that part of the valley could be seen downhill.

I looked for shotgun shells in the open, thinking maybe I'd get lucky and stumble across something that Jen and Brandon hadn't seen. When nothing came of that, I followed the trails that wind along and through the woods' edge, finding clearings where deer shelter, old-growth pines, and clusters of wild apples. The hawthorns had not dropped their orange-red fruit, and small birds foraged everywhere in the branches.

I saw no boot tracks but my own as I worked my way south and west, coming finally to an old canal and overgrown embankment that bisect the woods. The dike was good walking, and I followed it to where I could see the houses belonging to Brandon's suspects.

Looking through the leafless brush and trees, I watched the buildings for signs of life. Nothing stirred.

Weeks before, while out jogging after working at Schock's cornfield, I had turned up the lane to those houses. It was legal because the driveway is a public road. I don't know if the father or sons were there or what I'd have said if I met them. I only know that I was mad and wanted to confront the people who I thought had done so much wrong. I made it halfway up to the house before turning around.

Peeping from the woods like a wild creature, I wanted somebody to blame for what had happened to Millie and her cubs. The father and sons were obvious candidates, and like Brandon, I still thought it probable that one of them had pulled the trigger.

But looking up to where Laura's house sat at the base of the mountains, I couldn't be sure. Her house, barn, and outbuildings were far enough off to appear as a collection of small boxy shapes. A person walking around up there would appear as an ant-size upright slash. The naked eye could not see detail at that range, which undid Laura's account of seeing a tall thin man with glasses. Looking across the same four hundred yards, I did not believe that she or anyone could positively identify a person — even a familiar one — from so far off.

I came to a simple conclusion: I did not, and likely would not ever, know who had shot Millie and her cubs. Brandon had his theory, of course, and Laura her recollections, but nothing was certain.

The particulars of the sow's shooting were no better known than the culprit. Bears often cross paths with humans in the Mission Valley, and Millie was not the only grizzly to be shot at in the fall of 2016. There was, for example, the bird hunter who got charged, fired, and missed cleanly. He didn't mean to miss. For all I know, the only difference between that story and Millie's is the accuracy of a shot.

I know how natural it is to reach for a gun when the stakes are high. I used to work on a ranch not far from the edge of Yellowstone Park—a place that was lousy with grizzlies. When a handful of steers went missing, I headed out on a four-wheeler to search. Because I was afraid, I carried a .357 opposite the pepper spray on my belt.

Tick was with me, riding on the rear rack with his claws dug into the seat's Naugahyde. Miles passed, and we found nothing but empty forest and bolting elk. Deeper in the mountains, the roads dwindled to hard, slow going. In time, we left the machine and walked.

The carcass lay beside a brush-choked stream. A steer, two years old and weighing more than a thousand pounds, had been killed and partially eaten at the base of a pine. The remains were buried in a mound of duff, with the bloody head protruding at an angle from the top.

The place reeked of bear, a stink like that of spoiled meat and shit, and I could see the fur along my dog's back rising in a line. As he bristled, something stirred in the brush. I froze, expecting a charge. Tick stepped ahead, pointing like a compass needle. When the rustling noise moved, the dog did, too, cutting a tight circle around me.

It was only later, when Tick and I had backed away and were breathing again, that I saw the revolver in my hand, its hammer thumbed back on a round. I never blamed myself for that, because every creature takes the utmost care of its life and is right in doing so. I do not fault the bird hunter much, either, though I would put this question to him: Are there not better places to get pheasants? But asked that, he might answer honestly that there are no such places anymore, with the valley full of people and bears staying in the fields through autumn.

That hunter's world is changing. The ground is shifting under his feet, just as it shifts under the wide black paws of grizzlies in Schock's field.

An old rancher — a tribal member with sixty-some years of running cattle under his belt — told me recently that spring comes a month sooner to his pastures than it did when he was young. At first, I could not believe him.

"A month," he insisted, pronouncing the words softly and finally.

Gesturing down the valley at a clot of ranchettes, he told me that he hated houses.

What strains him and the bears is simple and evident: The climate is changing, and empty places are filling up. To use the parlance of grazing, our landscape is overstocked with people. More of us arrive — never mind our good intentions — every year, but the vessel that holds us cannot grow. So we chafe against one another or try to make more room by breaking land into smaller parcels. But a split farm, whether passed down to heirs or parceled out to strangers, seldom recovers. More often, it is halved out of being.

THE AIR was calm, and hearing the sound of cattle from the far side of the woods, I started in their direction. As I followed the lowing northward, the whole thing struck me as damn bleak. The sow died. Her cubs lost their freedom. The shooter got away, and the business seemed likely to be repeated next year.

On top of it all, the problem of Schock's field remained. Though my fence had done some good, Millie and several other bears had managed to cross it. With the bawling of dairy cows echoing through the trees, I wondered how much of the blame for the sow's ruin rested on Greg's shoulders. He, after all, had planted corn beside the mountains, attracting grizzlies and putting them in harm's way.

I could not consider him a villain for long. He is too good and too hardworking. Still, thinking of Greg left me worried, because the past months, starting with haying season, had changed him.

A madness takes ranchers in high summer. The sun is always up, virgin fields invite the sickle, and haymakers sprint to cut, bale, and stack enough feed to see their herds through winter. In their insatiable hunger to prepare for the cold, they are quite like bears.

All through the past August, I looked up from my work on the cornfield fence and saw Greg racing along the road, hauling spare parts and diesel from one place to another. He wore a set, harried expression through the busy season. That was all normal, but autumn should have freed him.

Every time we talked in November and December, Greg was further behind. He was always working on a broken-down machine and never had time to do more than read me his laundry list of troubles. Dairying didn't pencil out and hadn't for years. Milk was low, and he was thinking about getting out of the business. He asked me if I knew anyone interested in buying eighty acres of farmland.

"Be a shame to break it into twenties," he said. "But that's what I'm looking at, to get some money out of it."

Greg would do right by his land if he could. He is not the sort of man to overwork a cow, harm a grizzly, or wreck the valley with subdivisions if he can see a way to avoid it. As I moved slowly through the woods in the direction of Millie's old homestead, I thought that the problem was that Greg had been sold on a manner of farming that crushes people and wears out the ground.

Genetically modified seeds, synthetic fertilizer, diesel fuel, and herbicide: All of those things cost money and must be bought with the slim profits from selling milk and calves. The whole business is hard on the soil, wildlife, and a farmer's quality of life. It is unforgiving and inhumane, and I know that in Greg's place I could not stand it for long.

I like Greg Schock quite a lot. He is kindly and probably stronger than I am, but he makes his living on a vanishing margin against increasing odds. It is no wonder that he, like soil asked to raise corn for many years, is exhausted. As I considered this depletion, which is by no means unique to Greg or his fields, it seemed to me that he was as much injured by the cornfield as the bears.

THE WOODS have a dark and silent heart. It is offset from their geographic center, as the human heart is in the chest. The core of the woods, where few trails go and which I did not mean to enter, is south of where Millie's house once stood. Water comes up from the ground there, trickling away in slow streams. Birds move in the branches. A person must bend double or go on all fours through deadfall.

It was easy to blunder into the heart of the woods but hard to get out. As I pressed on, the seeps gave way to pools and impenetrable thickets. I had to navigate them, hopping between stones and exposed roots.

A large ponderosa stood on slightly higher ground, and coming closer, I saw that its trunk had been torn all over by grizzlies. There were parallel five-clawed rakes and head-high slashes. Smaller marks continued up fifteen or twenty feet.

For all the fearsome scenes that those gouges conjured — visions of growling and tearing and beasts that could shatter me with a cuff — the tree was lovely. Inspecting the scrapes, I found some fresh and others old enough to have grayed with the wood. Bears had been coming to the spot for a long time.

It was winter, monochrome and cold, but the marks on that pine held summer's promise. In warmer months, the woods' heart turns green and beats with life. It remains wet and cool through August when the rest of the valley is burning up.

Then the tree stands like a flagpole at the center of something. Because people, fearing bears, do not enter, animals live undisturbed. Boars visit the tree and claim it, giving one another wide berths. Sows pass cautiously, and cubs gambol in the shade. Deer

move through, always wary, with their semaphore tails working. Scrawny cats hunt voles and songbirds and are stalked in their turn by coyotes. The shadows of hawks slide over the canopy, sometimes dropping to the ground.

The tree held an echo of all this. It made me think of spring, though the new year had not come. Pulling off my glove, I traced a deep gouge with my thumb. Picking at the bark with my fingernails, I tried to leave a mark of my own. I could not do it.

Like any other modern man or woman, I am coddled by technology. I can climb into a tractor and stack a hundred tons of hay without sweating. If I pick up a rifle and aim well, a grizzly or any other beast will die. In a jetliner, snacking distractedly on pretzels, I can outstrip snow geese winging north. Such a life makes it easy to forget the softness of my hands and the fragility of my top-heavy body. But in the heart of the woods, I did not feel strong, quick, or even clever. I stood by the tree, small and naked of artifice, transfixed by deep gouges in the bark.

12

The
Exhibit

Aᴏ ᴀʟʟ their bad luck and staying much longer at the Helena facility than anyone expected, Millie's cubs caught a break. Around the end of November, I called Wayne Kasworm at the U.S. Fish and Wildlife Service. Dialing his number, I expected bad news or none, but picking up, Wayne sounded almost giddy on the line.

"A zoo in Baltimore," he said. "They're taking both of them."

The Maryland Zoo sits on a hill near the middle of a park that, in the spring of 2017, looked to me like an oasis. Snowdrifts were hanging on in Montana, but Druid Hill was green, with thick-trunked chestnuts and sycamores growing at intervals. Their branches spread black against the sky, with sunlight pouring through new leaves onto the grass. Some trees showed clustered purple flowers, their look reminding me how far I was from home.

I came to the park after three days in Manhattan, where I had felt restless and unable to keep still. Even at night, in my rented room, I crossed from window to door many times before forcing myself to sit on the bed. I never did relax in New York and always suspected it of tightening around me.

Driving into Druid Hill Park, I left that feeling behind. Inside the gates, a wide calm reservoir held the sky's pale blue. Joggers plied a path around it, and a muggy, windless heat gripped the day.

Crossing from shade into close-cropped meadows, the car climbed a rounded hill. A large white building crowned that knob, its roofline like a dollhouse's.

Michelle, the zoo's development director, had a crisp manner and an immaculate appearance that threw my traveler's shabbiness into relief. We rode a golf cart down to what she called the "campus," which was the fenced-in assortment of cages, pens, pools, and terrariums that constitute a zoo.

We did not go alone or directly to the bears. Michelle had arranged for me to meet and tour the grounds with one of the zoo's donors, a man named Bill who had contributed mightily to transport the cubs from Montana to Maryland. Bill was a substantial man of not less than sixty who operated from a position of childlike innocence. He was blithe, curious, and clumsy. How he made his money, I don't know. If we had been headed for the mountains or toward serious work, his manner would have worried me.

Waiting for the zoo's service gate to rattle open on its track, he talked at length about his relationship with one of the zoo's long-dead grizzlies.

"I'd go every day," he said. "And I'd bring an apple! He got to know me, you know? He'd watch me come to the rail, and I'd throw him his apple, and he'd *run* after it."

"Of course," Michelle said, "we don't let people feed the animals anymore. We haven't for a long time."

We held penguins. We peered through thick Plexiglas at lions while Michelle described all that had been done to keep the alpha male from impregnating one of his daughters. The giraffes were kept like horses in a tony stable, and they seemed pleased with their circumstances. I could not stand looking at the hopeless chimpanzees.

Parents and children walked the paths with us, and Bill tripped over a toddler. The boy cried. Both parties apologized. The day's heat built, and everyone but Michelle sweated.

"Wait until the school groups come through," she said, stopping us in a patch of shade. "Gets crazy."

THE ZOO'S bear exhibit encloses a space of approximately 100 by 160 feet, an area about half the size of my house's lot in the heart of Missoula. Because the enclosure was built with an arctic theme for polar bears, it is bisected by an ersatz tundra buggy — an off-road vehicle that looks somewhat like a jacked-up school bus — with gargantuan tires and tinted viewing windows.

A polar bear named Anoki has the eastern half of the display, a space dominated by a glass-walled pool and surrounded by concrete sculpted to look like sea cliffs. The sow was paddling when we arrived. Standing at the back of the gathered crowd, we peered through gaps between families.

She swam slowly, her long fur swirling hypnotically in the water. Insects trilled and buzzed in the trees, children pounded the glass, and she dove. Graceful and unhurried, she tucked, turned, and swam down ten feet after a scrap of food. Seeming as docile as a cow, she surfaced and chewed. One of the kids held up a white plush bear, shrilled "Anoki! Anoki! Anoki!" and pressed the toy against the glass.

Crossing to the buggy's far side, we found that the cubs had gone "off exhibit" for lunchtime. Their part of the display had a pond in the foreground, too, but it was smaller and without a see-through wall. Beyond the pond was a flat space. A few patches of grass remained, but the dirt was mostly beaten smooth. Beyond the flat was a hill terraced with boulders. It steepened toward the back, giving way to a concrete apron and a tall electric fence.

The public was out of luck, but Bill and I followed Michelle to the rear of the exhibit, through a locked door into a low building. Inside, it smelled like a calving barn or a country veterinarian's office — at once sour and sterile — and the bears' feeding times were posted on the walls. Refrigerators hummed in a corner.

Tanya, the keeper who showed us around, was outspoken and joyful. The bears seemed to be a source of joy in her life. Her way of talking about the cubs — familiar, interested, and emphatic, as though the bears were old friends of hers — went a long way toward putting me at ease.

"The big girl *loves* carrots," Tanya said. "The little one? Shy when she got here, but she's doing better now."

A rigorous discipline prevails within the exhibit, particularly in the corridor nearest the cages that house the bears when they leave the public eye. Foremost among the rules is this: No visitor may cross the yellow line painted at a two-foot distance from the edge of the pens, a line marking the farthest reach of the bears through their bars.

We followed Tanya into that corridor, which was murky and cavelike, with daylight entering through barred gates in the walls. Anoki had been brought in and was pacing a narrow run with her head held low. Her coat glowed softly and her eyes fixed on us with predatory constancy. The yellow line, and, indeed, the whole space we stood in, seemed well within her reach.

Bill didn't notice. Neither Tanya nor Michelle seemed bothered.

"What does the polar bear think," I asked, "about the cubs?"

"She's curious," Tanya said, glancing at the hulking, ghostly shape behind the bars. "Sometimes she acts like she's mom, and sometimes she wants to eat them."

She described how the sow would swipe at the cubs when separated from them by a single fence and how it frightened the little grizzlies. What the polar bear loved best, though, was exploring the cubs' side of the exhibit when they were out of it.

"Goes around smelling everything," Tanya said. "Keeps her busy for hours."

We walked toward a far corner of the building where daylight showed through an opening. A low shape filled that door, and Tanya spoke as we stopped at a partition of thick metal grating.

"Nova," she said. "Nova, c'mon."

The bear came without hesitation, moving out of silhouette into the building's gloom. Of what happened next, I remember fingers proffering a bright orange carrot; the soft and easy patter of words that Tanya maintained as the cub stood on her hind legs and pressed a paw against steel; claws curling into the space where we stood; pink tongue seeking, finding; teeth in fearful proximity to a hand; the panting breath that filled the room; the shining intelligence of the young bear's eye.

It was the larger of the two cubs, and she took carrots and apple slices with unexpected gentleness, mouthing them as carefully as she once must have taken her mother's teat. While she ate, I forgot about Anoki's pacing and the veterinary smell.

Soon, the smaller cub entered through the door's lit rectangle, keeping far off from us.

"C'mon, Nita," Tanya said, holding out a bit of food.

"Come, Nita. Nita, come," Tanya repeated evenly until both cubs were at the grating. She fed them at a distance from each other, walking back and forth to give each animal an equal share of food.

"Bill," she said, after moving back and forth for what could have been fifteen seconds or five minutes. "Watch your fingers."

Bill nodded, pulling his hands back from the grating and stuffing them into his pockets. Tanya returned to her work, praising the bears in a measured voice.

All was calm until the larger cub, following Tanya along the bars, got too close to the spot where her sister was feeding. In an instant, the grizzlies were at each other's throats. Locked in a clinch, they spun over the concrete floor, growling until the building rang. Though it was no more than a squabble over carrots that lasted just a moment, the fight impressed me. Some wildness remained in the cubs — and fury, too.

Their struggle caught the polar bear's attention. She stood at the near corner of her run, sniffing deeply, watching us all.

"The girls get into it," Tanya said. "But they love each other. They sleep in the den together, always. They have to touch."

"Bill," she continued without varying her tone. "You got to watch those fingers."

BACK ON the public side of the exhibit, Michelle and I stood together until a gate opened near the wheels of the polar buggy. The big cub appeared, shambling uphill for a few steps before coming down toward my side of the enclosure.

Emerging slowly, the smaller cub made straight for the boulders and began searching beneath them. She scraped at the ground, chewed, and swallowed. After watching her for a time, the larger cub began prowling similarly among the stones.

"Enrichment activities," Michelle said brightly. "We hide things all over the exhibit. Treats and toys. If they stay busy, it keeps them from getting stressed."

I asked how a person could tell when a bear was stressed.

"They start to pace. That's the number one sign. They'll pace all day or swim in circles for hours and hours. Anoki used to do it. Now, with the cubs, she's better. Like Tanya said, she's interested."

Michelle left me sitting on a bench in a young tree's shade ten yards from the waist-high glass wall overlooking the exhibit. I remained for a long time, and everyone who came to see the cubs passed in front of me.

I focused on watching, and the first thing I noticed was that the cubs didn't watch back. In Helena, we had made eye contact, with their gaze holding equal measures of fear and fascination. At the zoo, though, they couldn't have cared less that I was there. I stared straight at them for five, ten, fifteen minutes, thinking they would feel my attention, pause, and turn as animals do in the wild.

The cubs went on grubbing under rocks, and though they sometimes glanced in the direction of the crowd that flowed like a slow river between us, they took no notice of it. The cubs seemed listless, even lazy. When they were through finding hidden things, they lay side by side in the shade and waited. After a while, I got tired of watching them.

"They're tiny!" screeched a child. Her sunburned father, shaking his head and looking disgusted, said: "Sure, baby. Oughta be bigger. Grizzly bear's supposed to be big. Something wrong with these, maybe. Maybe that's why they got them here in the zoo."

Pausing by the exhibit, he shot a perfunctory glance at the bears and drank iced soda from a paper cup. The animals clearly bored him, and it stung me. *Ass,* I thought. *They look bigger without the fence.*

After that, I paid as much attention to the passersby as to the cubs. I learned that the average length of time spent watching the bears was between fifteen and thirty seconds and that most zoo visitors were content to look once and move along, saying the word "bear" or "grizzly bear" to one another.

The tree's small apron of shade crept away from my bench, and heat set me drowsing. People came and went at the exhibit's railing, and I listened.

A woman, pushing her son in a stroller, parking him front and center, said: "Honey, you got a bear just like that at home."

One teenage girl turned to another. "Made me sad — that thing in a cage."

A middle-aged lady with penciled-in eyebrows waved her hand as if brushing away flies. "Those ain't no polar bears. Look at them — they're brown. *Dirty* bears."

A frail white-haired woman shook her head at that, and said quite loudly: "Poor little orphans. Poor little *orphans*."

She stayed longer at the railing, standing for ten minutes in the sun, talking to a changing set of neighbors.

"They're young," she proclaimed. "But they'll get bigger. Big as that polar bear, maybe. One of them looks really good." She pointed to the larger cub. "Looks like a grizzly."

Tanya came up the path with a bucket in hand and tossed sliced fruits and vegetables to the bears. When she pitched a few bits into the pool, the larger cub took to the water. Paddling between apple slices and rolling otterlike to consume them, she was buffoonish and entertaining.

A knot of people gathered, and joining them beside the rail, I fell into conversation with a young blonde woman and her towheaded six-year-old son.

"I bring him here," she said, "so he'll learn to love animals. I wish he could see them in the wild, but this is the next best thing."

Her boy stared through the glass.

"The next generation," the woman continued, "has got to love animals."

The smaller bear refused to swim. After wading and retreating, she contented herself with hunting down pieces that Tanya had thrown among the boulders. She did this efficiently and with purpose, and when she had consumed everything, she padded to the door through which she had entered the exhibit. She stood with her nose near the bars, seeming so sure that the gate would open that I began expecting it to.

We waited. Somebody asked what the cub was doing and why it was just standing there.

"She wants out," I said.

The cub walked away along the fence to the pen's farthest corner and stayed there for a while. Coming back down, she stopped beside the gate before doing the circuit again. She moved like she had something in mind, and it made her look beyond the wires.

Giving up on the gate, she sprawled with her sister in the shade. Watching them, it was clear to me that the zoo had given Millie's daughters everything they could: food, shelter, and honest care. In spite of this, a cavernous lack remained.

Well-meaning care did not ease the sight of a paw pressed tightly to slatted metal, the claws curling through the gaps. It counted lightly against the fact that they could not go where they wanted to or choose anything for themselves.

Better off dead, I thought, but that wasn't true. I did not want them dead or kept. I wanted them foraging in the woods or haunting mountains where few people go. But the cubs would have none of that. Theirs would be the long, easy, withering road of captivity. Their muscles would never strain upslope. Their jaws would not close on living warmth. They would not walk nights through the valley, wake in dens to spring's stirring, or travel more than fifty steps in any direction. *All this follows from the sow's shooting*, I thought, as the bears waited side by side for the day to pass.

They ought to have meat, I thought in desperation. If never knowing a bear's life must be their lot, at least they should have wild meat. I would mail them a whole goddamn elk carcass — drive it down in a truck if I had to. They would have something from the mountains, I vowed. Next fall, when the weather cooled, I would find a way to send meat.

13

Near the Woods

When spring reached Montana, I returned to Schock's field to repair the winter's damage and ready the fence for another bear season. I worked hard at splicing wires, a low-to-the-ground job that left my knees aching and fingertips sore. It was the first hot day of the season, one that had started with frost and risen until the truck's thermometer topped seventy degrees.

Greg was plowing, making a second pass over disced ground. He drove with the tractor's loader arms lifted, so the machine appeared to have thrown up its hands in exasperation. The field looked wide, with sun-dried clods standing like whitecaps on water.

Kneeling, twisting a strand of high-tensile steel, I felt pain in hands grown soft over winter, then the pinging release of the wire giving way. The tractor drew a straight line of darker damp soil. Its noise grew, and I could feel dirt shuddering as the plow broke and turned it.

Looking up, I saw Greg near and high in the cab, crossing the same part of the field where Millie had been shot dead last fall. My recollection of the bear was distant, as though she had been turned under with the soil.

Every spring brings deathless days when winterkills are lost in a rising tide of grass. Birds wheel, courting. Early hatches plume above rivers, and the rich scents of growth and damp earth hang in the nose. The peaks are immaculate with snow, and it seems that they will never melt.

Greg pulled into the turn, bringing his slab-cheeked face into profile. Lifting a hand against the glare, I could see him clearly behind the wheel. Waving in salutation, he found his line, straightened his tires, and sent the machine across the expanse.

I watched him go, with my palm still covering the sun. He shifted often in the driver's seat, now looking back at his furrow, now ahead at the distance unplowed, turning the earth twelve feet at a pass and setting his year in motion.

I WANTED to jog in the late afternoon, so I drove uphill to the wide canal between the valley and the foothills, which is the source of Greg Schock's cornfield irrigation and the water sustaining most other farms thereabouts. After running north along the canal road, I realized that I had gone halfway through the upper end of Millie's Woods without considering the prospect of running into a grizzly. I blame springtime for that lapse.

Far downhill, Schock's field was visible as a patch of turned earth, a dark smudge on the greening landscape. Nearer was the uneven canopy of the woods. From where I stood, it stretched west like a muzzle sniffing at the low country. The aspens were leafing out, and young cottonwoods along the bank gave off a resinous smell.

The first track wasn't exceptionally big, but it wasn't small, either. A front paw's mark was recorded faithfully in mud, and my palm, when I set it down, was nearly as wide as the imprint. The track was fresh and still held all of its detail. Not twenty-four hours had passed since the last rain.

Now cautious and glimpsing open country ahead on the road, I ran past several more prints. Some pointed north along my route, and others crossed on a beeline for the woods. Several times ducks burst skyward from the canal, rising with a shattering sound.

It was a bad route, flanked by thick brush, but I kept to it until the woods opened to pasture. Clearing the timber, I stopped and panted with my hands on my knees.

Below me were the landmarks of Millie's last summer — the house where Brandon's suspects lived, the clearing where Laura had seen a man shooting at cubs, the body of the woods, and Schock's field. My thoughts ran backward, stuffing fire into a gun and making the sow whole.

She walked in the fields below the Missions, bedding along streams. She grazed in the cornfield, holding ears. Litters of cubs were with her, then gone. She became young and sleek, and the corn vanished from Schock's field. Meeting and following her own mother, she shrank until she was a cub denned in the pitch-black ice-bound haven of the mountains. I could have held Millie in my hand then. A careless step might have crushed her.

I stood on the ditch's bank, looking down at the valley's many houses, gridded roads, and well-used highway. Schock's field showed as a black speck, as small as a pip in a halved apple.

I have fenced the field, and fewer bears go into it, I thought. *So what?* I wanted very much to suppose that the sows and boars had returned to the mountains, seeking more natural sources of food. But looking across the lush wide Mission Valley, picking out orderly rows of fruit trees and the distant smudges of land plowed for oats, peas, and corn, I doubted that they would. There were too many other fields and orchards, too many lures that brought bears among the houses.

There are so many of us now. If we each do a little wrong, it will still be enough to consume and destroy what I hold dear. The population of the Flathead Reservation has increased 10 percent in fifteen years. Montana has grown similarly. Even here in what many consider an empty wild place, there are more people, houses, and roads.

If one person in a hundred leaves corn, chicken coops, or garbage cans unprotected, then takes a potshot at a sow come looking for a meal, it will be enough to damn the grizzlies persisting in Montana. Less — far less — callousness will suffice to check their movement south and the expansion of their range. Should this happen, a fence around Schock's field will have no more lasting effect than a match struck at midnight burning to its nub in the air.

There are good reasons for looking at things that way. In my lower moments, it seems that I could spend a lifetime building cornfield fences, worrying over cubs, and shipping elk meat to Maryland, and make no headway against our epidemic lack of restraint.

I AM vulnerable to despair when I think of what happened to Millie and how easy it is for other grizzlies to meet similar ends, but another story gives me hope. In 2015, a cub was born in the usual way, unseen in the higher reaches of the mountains. Brought downhill by his mother, he learned what a grizzly must to survive in the Mission. He saw trucks ripping along the roads, lit windows, and trash piles. He found out where the apple trees, dead pits, and cornfields were.

Whether he is any relation to Millie is unknown. He might well be, since she lived long enough to raise many cubs in the south valley. The boar could very well be her grandson or a nephew.

He was caught in a trap not meant for him. Fitting the bear with a GPS collar, Stacy named him Baptiste. Released in the foothills, he quickly moved south, as if fleeing. Skirting the Jocko Valley — a small pocket of farmed land south of the Mission — he detoured around houses and kept high up on the mountainside.

Crossing the Evaro divide, Baptiste descended toward the northwest corner of the Missoula Valley. He walked the cleared swath under high tension wires, leaving the Flathead Reservation for what would likely have been the first time in his life.

Coming through folded hills, he stopped on a bluff above the junction of Highway 93 and Interstate 90. He paused long enough at the forest's southernmost extent for his collar to record the location twice. From where he stood, he could have seen a corner of Missoula's sprawl. Houses spread out from the base of that hill, and roads snake off toward the city. Travel plazas mark the convergence of the big roads, their lights burning night and day. Beyond the sprawl are dark-green timbered hills — the beginning of a low range stretching south and west toward the Selway-Bitterroot Wilderness.

I cannot think of him standing there without urging him toward one of the narrow pathways remaining between the northern population of grizzlies and habitat farther to the south. Baptiste might have cut westward beyond Frenchtown to where the houses thin away. Out there he could have walked the bank of the Clark Fork River under Interstate 90 and been lost in the embracing mountains. He might as easily have struck east through the Rattlesnake Wilderness and reached the high valleys that stairstep up to Yellowstone.

But he was a bear already far from home, faced with a wide lake of human bustle. Turning away from the city, Baptiste began to retrace his steps north.

He spent the rest of summer eating fruit from trees on the edge of the Jocko Valley. Though the data from his collar shows that he lived among people — and a grizzly bear in the Jocko is uncommon enough to be news — he was not seen or reported. He survived the summer, and after ranging eastward to the Swan Valley, he denned below Gray Wolf Peak.

KNOWING HOW we have misused land and wildlife, I have precious little faith in humankind. I think it likely that we will go on wrecking the beautiful world. But I put my hope in bears of Baptiste's sort — hardy, seeking, adaptable creatures. They will find a way around or through our constructions to places that once belonged to them. Given the merest chance, they will live.

I will do what I can to give Baptiste his chance. While he slept through the depths of winter, Gillian and I bought a farm on the pretty uphill edge of the Jocko Valley. It is not as big as Schock's place, but it is a beautiful square of land with good soil tucked tightly beside mountains. Despite the fact that it was long ago cleared for agriculture and little brush remains for cover, Baptiste visited it often in late summer. He foraged at night for apples, coming within fifty yards of the house where Gillian and I will sleep. In the course of crossing fields where we will soon graze animals, he left silver-tipped fur in the barbwire.

We mean to root and make our living on that farm. We are committed. When Tick passed away with all of an old working dog's grace, we buried him there. Gillian and I will care for the fields where that good dog lies. We will be stewards of the place for as long as we are able, until the land outlives us. When we are gone, the farm will remain whole, open, and wild. It is now protected by a conservation easement so no future owner can subdivide it.

Baptiste, having visited the area and found it good, will almost certainly return to our corner of the Jocko. When he does, I will make sure that he finds no dangerous temptations. Our coops, fruit trees, and row crops will be ringed by electric fences. We will lock our garbage up tightly.

We will make room for him. For every acre fenced in or plowed, we will reserve at least as much space for Baptiste and his fellows. In the farm's southeast corner, several old ponderosas mark where forest once jutted out from the foothills. We will plant seedlings — pine, fir, larch, chokecherry, ninebark, Woods' rose, serviceberry — and let them grow until the farm's uphill edge is wild again.

ROUSING MYSELF and turning south, I started in the direction of my truck. Jogging back along the canal road into cool green timber, I peered into the shadows of Millie's Woods. Horsetail clouds had blown in from the southwest on a steady breeze that set the trees moving. No jays or songbirds called from the cottonwoods, and no mallards broke from the still water. There was only the noise of leaves and a feeling well remembered from my days beside the cornfield.

It was full spring, and bears had come into the woods. The place's entire aspect proclaimed this, so long-clawed tracks were hardly necessary to confirm it. Some of the grizzlies in there, I supposed, would be Millie's kin.

I wondered if I might recognize a creature of her line, then I gave up thinking. Bears waited in the tangled growth even if I could not see them.

I looked hard at shadows. Listening as I had learned to do beside the corn, I heard a charged and pregnant silence beneath the wind.

Sagging to the crest of the westward hills, the sun blazed orange. Heat fled the day. The change was sudden, as if my chest had been drenched in water. Night's chill was unmistakable, and the bears felt it. Rising and stirring, they prepared to go forth after gophers and new shoots.

Millie's offspring must have been among them. She had brought all her progeny to the woods, and those cubs, now fully grown, had not forgotten the sheltering trees. Perhaps some of them looked like the cubs in Baltimore, with pale bands of fur on their necks.

They moved, rising from shallow soft-earth daybeds, feeling

the dangerous mass of their bodies. They took deep drafts of air, inhaling the woods and everything upwind, smelling diesel exhaust, road dust, acres of turned soil, mice in holes, and trash cans at houses.

Bears stirred in thick cover, their coats becoming as dark as their shadows. Stretching forth long-clawed paws and crimping grass beneath the weight of leg, hump, and neck, they walked. Loose-lipped heads swung with the motion. Bright eyes moved ceaselessly. Aspens shivered above them, catching and scattering the last light.

Except for the resonant huffing of their nostrils, the grizzlies were silent. I never heard or saw them and kept to the road's grassy center as the westward sky reddened and the east bruised. We moved carefully in the twilight, navigating our perilous world, all bound for the edge of the woods.

AUTHOR'S NOTE

Telling a true story is a delicate business, particularly if the writer cares about his characters and wants to go on working with them once his book is in the world. I respect many of the people who appear in this story, so I have done my best to render them faithfully. Where necessary, as in the case of suspects and witnesses, I have changed names and details.

Much of this book was written from experience. Where the narrative moved beyond my life — into the realm of bear biology, for instance, or the details of a federal investigation — I researched and pestered experts to get the details right. I made particular use of several papers by Christopher Servheen, including "Grizzly Bear Dens and Denning Activity in the Mission and Rattlesnake Mountains, Montana," "Grizzly Bear Food Habits, Movements, and Habitat Selection in the Mission Mountains," and "Grizzly Bear Ecology and Management in the Mission Mountains." Richard Mace and Charles Jonkel's paper, "Local Food Habits of the Grizzly Bear in Montana," was also useful, as was Robert Klaver et al.'s "Grizzly Bears, Insects, and People: Bear Management in the McDonald Peak Region, Montana." In writing about the history of the Flathead Indian Reservation and its inhabitants, I drew from conversations with tribal members and consulted the Salish-Pend d'Oreille Culture Committee's "A Brief History of the Salish and Pend d'Oreille Tribes."

The book's dialogue comes from memory, notes, and a series of recorded interviews with the major characters. Throughout, I have tried to reconstruct conversations accurately without robbing the narrative of clarity, momentum, and grace.

Essential information about the movements of Millie, Baptiste, and other grizzlies in the Mission Valley comes from radio-collar data collected and shared by Stacy Courville on behalf of the Confederated Salish and Kootenai Tribes.

That data goes only partway toward rendering a bear's life. To get the rest of the picture, I drew from an understanding of a grizzly's seasonal rounds, my own experiences from a decade spent ranching in the mountains of Montana, and the accounts of the people who knew the most about Millie's life and death.

Because I could not walk with Millie, I hiked alone on similar paths in the Missions. Because I could not peer inside her den to see the cubs, I climbed into other unoccupied holes. Because I was not present to see how she ended up full of bird shot, I talked to everyone who had a theory about it, learned all I could, then wrote the scene as I thought it likely to have unfolded. If I have erred in the details, perhaps the shooter will come clean and set me straight.

Down from the Mountain is braided from research, experience, and invention. All three strands were necessary to make the story whole.

July 18, 2018
Arlee, Montana

ACKNOWLEDGMENTS

I am grateful to the bears of Millie's Woods for not eating me. They showed uncommon restraint — or perhaps were too full of corn to stomach anything more. In any case, I appreciate it.

I thank my good dog Tick, now departed, for standing lookout over the course of the last eleven years. He was attentive, vigorous, and mostly stalwart in the face of danger. I especially miss his company when I work in grizzly country.

My fiancée, Gillian Thornton, read nearly every draft of this story. When I was mired in the doldrums of writing and editing, we talked things through. When I needed help in the field, she put on work gloves and toiled. Her grammar is better than mine, and her heart is kind. For the disparate pieces of my life — farming, writing, bear wrangling, hijinks, love, etc. — I could have no better partner.

My parents, Colleen Chartier and Richard Andrews, raised me to look creatively at everyday events and perceive beauty in light and darkness. They set me on the path that leads to books like *Down from the Mountain.* I am particularly grateful to my mother for noticing, proclaiming, and insisting upon the individuality and value of animal life. I learned a lesson from her engagement with Millie's story and her reaction to Millie's death.

I discussed this project extensively with two writer friends: David James Duncan and Michael Hicks. David was generous with time, correspondence, advice, and varieties of Scotch that I cannot afford. His curiosity helped convince me of this story's worth. Michael did me the favor of approaching *Down from the Mountain* as if he had written it. We talked about minutiae, parsed details, and generally supported each other in the labor of creating prose worth reading. Michael also helped on the cornfield fence. We set posts together, though I have lately noticed that certain of them have not remained tight. All the same, I was glad to have his help.

My agent, Duvall Osteen, undertook the task of finding a home for this book, making a fraught process seem easy. Naomi Gibbs provided that home at Houghton Mifflin Harcourt and thereafter labored to refine and improve the manuscript. I want to thank her and others, including my copy editor, David Hough, for lavishing care and attention on my story. It would be less coherent without them.

My sincere thanks to Stacy Courville, Shannon Clairmont, Brandon Couture, and other employees of the Confederated Salish and Kootenai Tribes for sharing data about Millie and other grizzlies. This story would not exist without their help, and grizzlies would not be thriving in the Mission Range without the diligent, long-term efforts of the tribes.

Greg Schock let me experiment in his fields, thereby setting this story in motion. He has been patient with the bears and friendly toward me. I thank and commend him for it. Finally, I want to recognize Steve Primm, Lisa Upson, and my other colleagues at People and Carnivores for supporting the work that led to *Down from the Mountain*.

A CONVERSATION WITH
BRYCE ANDREWS

You encountered Millie, the grizzly bear at the heart of *Down from the Mountain,* while doing conservation work in the Mission Valley in Montana through the nonprofit People and Carnivores. How did you become involved in this sort of work?

I've loved wild creatures and remote places since I was a child. That deep-seated interest drove me east from Seattle, where I was raised, to western Montana, where I made my living as a ranch hand for a decade. I saw a lot of wolves and bears on ranches. They caused trouble, but they also intrigued me. Something about predators cuts straight to our center, stripping away distractions and pretense. I value that, and want to make sure that space remains for such animals in the modern geographical and cultural landscape.

Ranching brought me close to wolves and grizzlies, even when I worked against them. That intimacy helped me to recognize the importance of those species. It inspired me to work on their behalf. When I connected with People and Carnivores, a group approaching conservation with a deep respect for the people living in rural, wild places, it was a natural fit. They offered me a job and I took it.

What drew you to ranching in the first place?

It started with an art exhibition and a road trip. My father was the director of the Henry Art Gallery in Seattle and he organized a show called *The Myth of the West.* I was young and the artwork hooked me. I remember looking at images of Montana — of Yellowstone and herds moving under endless skies — and thinking that I had to go and see it. We took a car trip out to Montana in the summer, and the place was bigger, lovelier, and rougher than I could have imagined. My parents had friends, Pat and Suzie Zentz, who ranched outside of Billings. Pat was particularly important to me because he was a rancher and an artist. He presented the work of agriculture as a method of looking deeply at land. He led me to believe that the task of making a living from the soil, if approached with the right sort of attention, could be endlessly complex and meaningful. I still think that's true.

Pat and Suzie were part of what drew me to ranching. The Sun Ranch was equally important. My first book, *Badluck Way,* came from the Sun. I fell into a summer job at that place, which sat at the higher, colder end of the Madison Valley, just outside Yellowstone Park.

There were wild animals everywhere — huge herds of elk, packs of wolves, roaming bears — and I had a rare and intimate view of it all. The job convinced me that ranching would keep my life interesting.

How does Millie's story speak to broader issues at play in the American West? How does this particular narrative fit into a larger one in the region?

Millie's story opens onto many of the modern West's essential issues. She's an entirely wild animal who must make her living at the edge of a domesticated landscape. She encounters rural sprawl, speeding vehicles, herds of untended livestock, and human beings with trigger itch — in other words, the West. She gorges on corn grown from genetically modified seed and digs daybeds in glyphosate-soaked soil. Raiding trash cans, she ends up eating the same processed, overpackaged garbage that we do. Her struggle to survive unfolds against the backdrop of a warming climate, with natural food sources dwindling, residential development fragmenting bear habitats, and irresistible crops being planted on the valley floor.

The particulars of Millie's life and death show clearly what we've done with Montana's pretty valleys. I hope her story makes an argument for changing things.

Do you think having worked as both a rancher and in wildlife conservation gives you a unique perspective on the issues running through this book?

I do. Having spent a good deal of time in the saddle, fixing fences, and otherwise doing the mundane tasks of ranching, I've developed a profound respect for agriculture and the people who pursue it. I know the work is hard — harder than most jobs in the modern economy. I understand the ways in which it can be rewarding and devastating, and how the presence of wolves and bears can tip a delicate balance in a painful direction. Put plainly, I understand how tough it is to make a living from the land.

I'm also aware that our approach to agriculture needs to change. Having shipped hundreds of calves to slaughter and grazed cattle on thousands of acres of private and public land, I understand that we've come to a point of reckoning in the West.

Many ranchers have cast their lot with scale and efficiency, grow-ing their operations by taking on debt and embracing things like genetically modified seed, animal growth hormones, and exten-sive herbicide use. One look at the number of family ranches going broke, or the staggering acreage lost to subdivision and ir-responsible development, is enough to show that this approach isn't working. We need to do better by the land that sustains us. We all have a right to push for this, as well as a responsibility to check ourselves and others who fall short.

There's a lot of tension in this book and clearly a lot of conflict in the West over whether or not we should be protecting grizzlies. Why do you think this work is important? Are there any efforts in particular that give you hope?

Coexisting with big carnivores always generates conflict and tension. It's a difficult subject, because when we argue about our responsibilities and approach to living with grizzlies, we're really talking about whether humankind should inconvenience itself on behalf of other species. Should we slow down on the high-way? Ought we refrain from raising livestock in certain places? Should we require communities to forgo growth or prosperity to preserve habitats for dangerous beasts? These are not simple questions.

Grizzly bears are hard to live with, and getting along with them requires restraint and care. Those two qualities are com-mon enough in individual humans, but they're rare in groups and societies. Fostering those virtues is a big part of my work for People and Carnivores.

Certain things give me hope. Across Montana and the West, more people are recognizing the cultural, economic, and biological value of species like wolves and grizzlies. We're beginning to get creative about coexisting with large predators.

That curiosity takes many forms, but livestock guardian dogs are particularly interesting to me right now. Big, tough dogs are used all over the world to protect livestock. They've been bred to that purpose for millennia and are very good at their job. I'd like to see a lot more dogs protecting herds and flocks in the West. I've been working on several such projects and will experiment in the future with using dogs to keep grizzlies out of crop fields.

Perhaps most encouraging is the way conservation groups, governmental agencies, and agriculturalists are beginning to look more holistically at landscapes. By asking the right questions and making use of information from collared wildlife, we're coming to better understand how species move through Montana's patchwork quilt of wilderness and development. We know, for instance, of a grizzly bear that makes an annual seventy-mile trek, crossing three valleys to arrive at a farmer's field when the corn is ripe. Understanding such connections allows us to be proactive in our efforts to limit conflict. Right now, we're maturing into a better understanding of grizzlies and a clearer sense of what they mean to us. The next essential task will be acting on that knowledge.

QUESTIONS FOR DISCUSSION

1. In what ways has modernity threatened wildlife in Montana? What are some ways in which you've observed humanity's effects on wildlife where you live?

2. How does Andrews understand the agency and intellect of animals? What wisdom did he gain from observing and working with animals as a rancher, and how does he use his understanding of domesticated animals to help him understand wildlife — in particular, carnivores?

3. On page 25, Andrews writes the following about ranching: "Of this I can be certain: I recognized a void and a canker, and I was certain that it had something to do with killing. I suspected, too, that making amends would require saving the lives of animals." Discuss what Andrews means by "a void and a canker" and the emotional tolls of ranching.

4. What does success look like for animal conservationists like Stacy? Why do we need experts like him, and what are the limitations of their work?

5. Stacy expresses his frustration with hunters' "lack of concern" when it comes to grizzly bears, explaining "how easy it was for a grizzly and a hunter to surprise each other — an

encounter that often led to a frightened gunshot" (155). Discuss the relationships and power dynamics between hunters, grizzlies, conservationists, and ranchers. Compare and contrast the ways they each live off the land. What happens when their needs for space or resources come into conflict?

6. What is the impact of Association of Zoos & Aquariums–certified facilities and rehabilitation centers on local wildlife? How does AZA certification facilitate or impede the rescue of certain animals?

7. Lisa, the manager of the WILD Wildlife Rehabilitation Center, observes: "Zoos aren't doing too well these days . . . Attendance and funding have flagged as the public grows less comfortable seeing animals in cages" (205). How do you feel about the decline in zoo's popularity? What are some positive and negative impacts of zoos?

8. What responsibilities does Andrews take on for the animals in his life, and how do they shift throughout the book? How has reading this book changed the way you see animals and human-animal relationships?

9. Andrews was not raised in Montana. What are some key moments in which he shows us how his passion and respect developed for the place, its inhabitants, and its wildlife?

10. How are organizations like People and Carnivores approaching entire ecosystems holistically? How has technology aided in restoring the populations of wolves, grizzlies, and other wildlife? What, to you, might true coexistence look like?